ZEW Economic Studies

Publication Series of the Centre for European Economic Research (ZEW), Mannheim, Germany

ZEW Economic Studies

Vol. 1: O. Hohmeyer, K. Rennings (Eds.)
Man-Made Climate Change
Economic Aspects and Policy Options
1999. VIII, 401 pp. ISBN 3-7908-1146-7

Vol. 2: Th. Büttner
Agglomeration, Growth, and Adjustment
A Theoretical and Empirical Study
of Regional Labor Markets in Germany
1999. XI, 206 pp. ISBN 3-7908-1160-2

Vol. 3: P. Capros et al.
Climate Technology Strategies 1
Controlling Greenhouse Gases.
Policy and Technology Options
1999. XVIII, 365 pp. ISBN 3-7908-1229-3

Vol. 4: P. Capros et al.
Climate Technology Strategies 2
The Macro-Economic Cost and Benefit
of Reducing Greenhouse Gas Emissions
in the European Union
1999. XIII, 224 pp. ISBN 3-7908-1230-7

Vol. 5: P.A. Puhani
Evaluating Active Labour Market Policies
Empirical Evidence for Poland During Transition
1999. XVI, 239 pp. ISBN 3-7908-1234-X

Vol. 6: B. Fitzenberger
Wages and Employment Across Skill Groups
An Analysis for West Germany
1999. XII, 251 pp. ISBN 3-7908-1235-8

Vol. 7: K. Rennings et al. (Eds.)
Social Costs and Sustainable Mobility
Strategies and Experiences in Europe
and the United States
1999. VI, 212 pp. ISBN 3-7908-1260-9

Vol. 8: H. Legler et al.
Germany's Technological Performance
2000. X, 191 pp. ISBN 3-7908-1281-1

Vol. 9: O. Bürgel
The Internationalization of British Start-up Companies in High-Technology Industries
2000. XIV, 230 pp. ISBN 3-7908-1292-7

Vol. 10: J. Hemmelskamp et al. (Eds.)
Innovation-Oriented Environmental Regulation
Theoretical Approaches and Empirical Analysis
2000. VI, 348 pp. ISBN 3-7908-1313-3

Karl Ludwig Brockmann · Marcus Stronzik (Eds.)

Flexible Mechanisms for an Efficient Climate Policy

Cost Saving Policies and Business Opportunities

Proceedings of an International Conference
held at Stuttgart, Germany, July 27–28, 1999

With 23 Figures
and 6 Tables

Springer-Verlag Berlin Heidelberg GmbH

ZEW
Zentrum für Europäische Wirtschaftsforschung GmbH
Centre for European Economic Research

Series Editor
Prof. Dr. Wolfgang Franz

Editors
Dr. Karl Ludwig Brockmann
Dipl.-Wirtschaftsing. Marcus Stronzik
Centre for European Economic Research (ZEW)
L 7,1
68161 Mannheim
Germany

ISBN 978-3-7908-1314-2

Cataloging-in-Publication Data applied for
Die Deutsche Bibliothek – CIP-Einheitsaufnahme
Flexible Mechanisms for an Efficient Climate Policy / ZEW, Zentrum für Europäische Wirtschaftsforschung GmbH. Karl Ludwig Brockmann; Marcus Stronzik, ed. – Heidelberg; New York: Physica-Verl., 2000
(ZEW economic studies; Vol. 11)
ISBN 978-3-7908-1314-2 ISBN 978-3-642-57691-1 (eBook)
DOI 10.1007/978-3-642-57691-1

Originally published by Physica-Verlag Heidelberg in 2000

Cover design: Erich Dichiser, ZEW, Mannheim

SPIN 10771352 88/2202-5 4 3 2 1 0 – Printed on acid-free paper

Preface

We all hope that the ratification and implementation of the Kyoto Protocol will soon be achieved. This protocol will be the first legally binding instrument to reduce greenhouse gas emissions worldwide. A constructive dialogue on the flexible mechanisms and their environmentally and economically effective design is of great importance.

Compared to other climate policy instruments, flexible mechanisms are not only market based and cost efficient, but can also support the complete attainment of a given ecological target like the Protocol's emission limitations. In addition, they adjust to changing economic conditions, allow for an application at the international level, and do not cause excessive administrative costs.

The aim of our conference on "Flexible Mechanisms for an Efficient Climate Policy" which was held in the capital of the German State of Baden-Württemberg in July 1999, was twofold. Firstly, the purpose was to make politicians, business executives and the general public more aware of the need for, and the potentials of, cost efficient climate policy instruments. Secondly, it offered a forum for economists and lawyers to discuss the economic and legal aspects of flexible mechanisms – international emissions trading, joint implementation and the clean development mechanism.

The conference was jointly organized by the Centre for European Economic Research (ZEW), Mannheim/Germany, and the Ministry for the Environment and Transportation of the State of Baden-Württemberg. This volume contains most of the contributions presented by politicians, scientists, and business executives. They cover a broad range of issues which are and will be important for the design of a climate policy which is both effective in reaching the given targets and efficient by avoiding unnecessary costs for consumers, companies and economies.

We hope that some of the ideas discussed at the conference and in these proceedings will enrich future discussions at the UNFCCC conferences in Bonn and the Hague, as well as in Berlin, Stuttgart and other capitals of German Länder.

Stuttgart, March 2000

Ulrich Müller
Minister for the Environment and Transport of the State of Baden-Württemberg

Contents

Introduction

Karl Ludwig Brockmann and Marcus Stronzik

Centre for European Economic Research, Department of Environmental and Resource Economics, Environmental Management, P.O. Box 103443, D-68161 Mannheim, Germany, e-mail: brockmann@zew.de and stronzik@zew.de.

Introductory Talks

The introductory talks of the conference were started with the urgent demand by the Minister for Environment and Transport of the German State of Baden-Württemberg, Ulrich Müller,* for a new policy style in German and European climate policy. Natural science should set the ecological targets, and economic science should tell us which instruments to use.

Wolfgang Franz, President of the Centre for European Economic Research (ZEW), Mannheim/Germany, supported this by pointing out that cost efficiency in climate policy is necessary for the well-being of any economy that is exposed to a growing international competition. In addition, a policy that makes use of efficient climate policy instruments allowing for a flexible compliance with the given ecological targets not only helps avoiding unnecessary costs for the economy. In addition, it helps to integrate the developing countries into international climate policy, and raises the acceptance for urgent new commitments to reduce emissions of greenhouse gases beyond the Kyoto targets.

---

* This contribution – as much as the contributions from Wolfgang Franz, Franz Josef Radermacher, Tobias F.N. Schmidt, Christine Zumkeller and Heidi Bergmann, all mentioned in the following – is not contained in the proceedings. In addition, we incorporated the original contribution from Viktor Danilov-Danilian, although at the conference he was represented by Sergey N. Kuraev. Finally, at the conference Peter Zapfel gave an overview of European climate policy, whereas his contribution to these proceedings tackles only one specific issue, the use of the Kyoto mechanisms as a supplement to negotiated agreements.

Stefan Rahmstorf from the Potsdam Institute for Climate Impact Research, Potsdam/Germany, presented recent research on climate risks from anthropogenic greenhouse gas emissions: A major flaw in thinking about global warming is to interpret the global climate as a linear system. Instead, it is a non-linear system, where at some critical point, i.e. at some global temperature rise, the system can transform very quickly into a totally different state. For example, the Atlantic system of currents is not entirely stable and important currents might stop. This would shut down Europe's central heating and would lead to a drop in temperatures in Europe. Where exactly these points lie, and the behavior of the climate near such points, is difficult to calculate. For the way we are dealing with the climatic system, the new findings mean, above all, more insecurity. The more we drive climate away from its present balance, the higher the risk of disagreeable surprises becomes.

Hans Jürgen Ewers, member of the German Council of Environmental Advisors (SRU), stressed that the physical, cultural and economic setting is different from country to country, which has severe consequences for the choice of instruments in climate policy. Nevertheless, for any country the implementation of economic instruments in climate policy should be an option worth being considered. On the hot air topic, he took the clear stance that this is not a problem relating to the instrumental choice. Whatever level of flexibilization by emissions trading or joint implementation might be chosen, the overall target of reducing greenhouse gas emissions in Annex B countries by 5.2% will be met. In the end, any criticism about hot air is criticism about the targets and not about the instruments.

Franz Josef Radermacher, Director of the Research Institute for Applied Knowledge Processing in Ulm/Germany, stressed that developing countries should enjoy the same rights to emit greenhouse gases as the countries of the developed world. Hence, he supported the determination of the same amount of per capita emissions of greenhouse gases for all human beings: "We will not reach a comprehensive solution of the climate problem as long as the South may not catch up". The way to a sustainable world-wide information society which overcomes distances, must also be a sustainable path for world development. Growing welfare must be accompanied by a significant rise in efficiency. At the same time, policy and ethics have to set clear boundaries to the domain of the markets.

Social Costs of Climate Protection

Economic models can give us quantitative estimates on how much we can gain from implementing flexible instruments – which allow the exploitation of cost-differences in abatement across different emission sources – in national and international climate policy. Christoph Böhringer from ZEW showed that potential welfare losses can be reduced significantly by the use of these instruments. This is also true for developing countries which are not yet subject to a binding cap on their greenhouse gas emissions under the Kyoto Protocol, as negative international spillover effects from the abating developed world to the non-abating developing world would be reduced substantially. Thus, flexible mechanisms such as emissions trading or joint implementation seem to be necessary in order to include developing countries in a post-Kyoto strategy. In addition, the economic gains that go along with flexible instruments considerably relax the problem of fair burden sharing across regions and make it easier to comply with basic equity criteria.

In his model, Tobias F.N. Schmidt from ZEW revealed that the allocation mode of grandfathering emissions rights in a greenhouse gas emissions trading scheme can be of importance if general equilibrium effects are taken into consideration. If a country joins an international emissions trading scheme and becomes a net seller of emissions rights, this might result in a net welfare loss to the country. This finding, which is a little bit counterintuitive, is due to the differences in the calculus of an emitter holding emission rights, and the calculus of a benevolent welfare maximizer. To the emitter, e.g. a great power producer, it might be profitable to invest in new efficient electricity production equipment and to sell emission rights, if he finds that the prices of emissions rights in the international market are relatively high. But, he would not care for the economic losses his home country may suffer from rising electricity prices. A maximizer of overall welfare would invest less in order to keep prices low.

Kyoto Mechanisms

Christine Zumkeller, with the United Nations Climate Change Convention Secretariat, pointed out the fact that, considering the projected rise in greenhouse gas emissions, the quantified emission limitation agreed upon in the Kyoto Protocol – though not yet ratified – is by no means a negligible target. But, without additional measures the Kyoto targets will be missed, and the effective implementation of the Kyoto mechanisms depends decisively on the capacity building efforts of industrialized and developing countries. All in all, the post-Kyoto process appears to be more promising than a year ago.

Nathalie Eddy, US Climate Action Network (USCAN), Washington, D.C., evaluated the international climate policy from a non-governmental organization's (NGO's) perspective – CAN is the umbrella organization of environmental NGOs in the international climate negotiations. She stressed that national climate policy measures should take precedence over international flexible measures like international emissions trading, as the necessary turn in global greenhouse gas emissions can only be reached by national measures. Any approach to international flexible measures should set clear and comprehensive rules in order to avoid ecological loopholes. The project-based mechanisms joint implementation (JI) and clean development mechanism (CDM) raise the question of additionality; therefore, only projects screened by a strict verification and certification process and based on "clean" best available technologies should be acknowledged.

Both project-based mechanisms according to Art. 6 (JI) and Art. 12 (CDM) of the Kyoto Protocol were also discussed by Axel Michaelowa, Hamburg Institute for International Economics – HWWA, Hamburg/Germany. Whereas credits from CDM-projects can be gained from 2000 on, those from JI are not available until 2008. Another important difference between the two mechanisms lies in the recipient countries, which are Annex I countries to the Climate Framework Convention for JI, and non-Annex I developing countries – for which no binding national cap exists – for the CDM. This would mean, Michaelowa argued, that special attention must be paid to the emission reductions gained from CDM-projects in order to avoid the production of emission credits, which do not represent real emission reductions compared to business as usual. One major disadvantage the CDM has to face is that projects under Art. 12 are subject to an adaptation and administration tax whereas JI is not. Recent studies reveal that, compared to international emissions trading (IET) and JI, the CDM bears a high potential. This is of particular importance from a development perspective. One of the most frequently discussed feature of JI and CDM is the baseline constructed to compute emission reductions gained. Here, Michaelowa supported an institutional competition between alternative approaches.

Viktor Danilov-Danilian from the Russian Federation State Committee on Environmental Protection stresses in his paper that one important challenge with flexible mechanisms is to create procedures for the verification of compliance, and for the determination of financial burdens for all participating countries. As these procedures must be objective, regular, convincing, and fair market procedures, instruments like greenhouse gas emission trading are needed. Carbon taxes will ensure compliance with agreed emission limits only if a complex feedback mechanism – requiring an immense concerted international bureaucratic effort – is established. Whereas a strong feature of emission trading is its decentralized character taking considerably less administration costs.

Real world experience with permits trading, presented by Garth Edward from the Natsource Emissions Brokerage Desk in New York, is promising. US trading schemes for sulfur- and NO_x-emissions helped reduce costs and stimulate technical and organizational innovation. In addition, there is already a small but growing greenhouse gas market, e.g. from international initiatives, from many countries' early action pilot trading programs and crediting legislation, and from US State voluntary greenhouse gas initiatives.

Implementing Flexible Mechanisms

Gerd Svendsen from the School of Business, Aarhus/Denmark, offered insights into the political economy of instrumental choice. He showed that political decision makers may not have clear incentives to choose the most efficient environmental instruments. Considering the different abilities of addressees to organize and lobby, a mix of green taxes (in relation to non-organized interests) and grandfathered permit markets (in relation to organized interests) should be considered in the search for cost-effective and politically feasible instruments.

From the legal analysis of Heidi Bergmann, ZEW, we learned that, although it is in principle relatively easy to implement flexible mechanisms into the German and European Legal System, problems may arise in questions of detail. Emissions trading approaches are fundamentally different from the way the European Community (EC) and its Member States have organized their environmental policy over the last decades, as these were characterized mainly by technical standards and emission limits, rarely by taxes and charges, or by the more recent voluntary agreements. Nevertheless, operating with flexible mechanisms would mean a "switch of philosophy" but not a breach of basic law principles. It is possible for the Member States to undertake actions on their own initiative. The EC, however, has to ensure that measures being taken at both the Member State and Community level are consistent with other Community policies and that they respect the Treaty. For internal market reasons, it would appear to be preferable to consider an EC-wide permit market.

Peter Zapfel from the European Commission emphasized that although the integration of emissions trading into European Climate policy is a considerable challenge, it is worth facing. In his paper he investigates the combination of flexible mechanisms with negotiated agreements. The European implementation debate to date is characterized by a strong preference expressed by industry stakeholders to use these negotiated agreements on energy efficiency as the primary instrument to achieve the sector's contribution to the Community objective. An analysis of several models for combining negotiated agreements with flexible mechanisms in the industrial sector shows that it is not possible to develop a policy mix that provides for a high degree of ex-ante quantity certainty for the regulator, while avoiding absolute emissions limitations for affected industries.

The presentation of Peter Knoedel, BP Amoco, made clear that emissions trading can be an important tool for a private company, and that companies themselves can engage in climate protection without "hard" incentives by legal environmental instruments. Climate change may be increasingly seen as a competitive issue among companies as it is very likely that the issue will become more significant for governments and the public.

Finally, Annie Petsonk, Environmental Defense Fund, Washington, D.C., and Ray Kopp, Resources for the Future, Washington, D.C., took two different perspectives on options for the United States to act early, i.e. to implement climate policy instruments before the budget period of the Kyoto Protocol will begin in 2008. Whereas Annie Petsonk described a framework for voluntary early action which would create credits before 2008 that companies could use after 2008, Ray Kopp proposed a comprehensive and mandatory emissions trading scheme which could start in, e.g., 2002 and would establish a generous cap on greenhouse gas emissions and set an upper price limit on emission rights, in order to avoid disruptions in the economy. This approach is a combined tax-permit scheme.

All in all, it appears that flexible instruments can considerably lower the costs of climate protection not only for emission sources but also for the economy and for economies linked to countries with a high emission volume. After all, real world experience shows that market-based instruments are a promising efficient tool to tackle the environmental problem of climate change. But when implementing flexible mechanisms – be it before or during the budget period – it will be necessary to take care of possible loopholes and of uncertainties and transaction costs which may be too high if no proper design is chosen.

Anthropogenic Climate Change: The Risk of Unpleasant Surprises

Stefan Rahmstorf

Potsdam Institute for Climate Impact Research (PIK), P.O. Box 60 12 03, D-14412 Potsdam, Germany, e-mail: rahmstorf@pik-potsdam.de.

During the past few years, our understanding of the global climate has gone through a silent revolution. Our old image was based on slow climatic cycles, in which over many millennia ice ages came and passed, driven by the slow cyclical changes of the Earth's orbit around the sun.

This concept was, above all, derived by retrieving cores from the muddy sediment layers at the bottom of the oceans, which were deposited over the course of many millions of years and form a unique climatic archive. Since in most places only a few centimeters of sediment are deposited over a thousand years, these cores revealed at first only the slow climatic cycles. It was possible, for example, to make out that the last ice age started about 120,000 years ago and ended 10,000 years ago. During this glacial period our ancestors had to survive by hunting mammoths in the icy steppes. The stable period of warmth which followed (and which prevails until today), the Holocene, was the prerequisite for the development of agriculture and modern human civilization.

Climatic Jumps and Jolts

A completely new picture of climatic history was gained mainly from ice cores drilled from the Greenland ice cap, which allow a much higher resolution in time (Blunier *et al.* 1998). In these ice cores it is even possible to distinguish the individual annual layers of winter snowfall, similar to the annual rings of trees. In addition to the slow cycles already known, abrupt climatic changes could now be identified, during which climatic conditions changed dramatically within a few years. At the end of the last glacial period, when the planet was generally warming up, a sudden lapse into an extreme period of cold occurred in the Northern Hemisphere, the so-called Younger Dryas event, which lasted for almost one thousand years. During the last ice age, altogether 24 such abrupt cooling events are recorded. Clearly the climatic system has the tendency to make sudden jolts.

The causes of rapid climate shifts have not been fully understood so far. However, everything points to the fact that they are not caused by sudden changes in external factors like solar radiation, but lie in the changeable nature of the climate system itself. In other words, climate is a strongly non-linear system. Linear systems are very simple: if they are forced, they react more strongly the greater the forcing, or stimulus, is. The connection between stimulus and reaction is a straight line – it is "linear". Complex systems are almost always strongly non-linear. These are characterized by a tendency towards self-regulation and sudden transition to a qualitatively different state when a certain critical threshold is exceeded.

Swimmers in the Sahara

The climate system appears to have the ability to self-regulation, within certain limits. An interesting example of this is the Sahara. Alexander von Humboldt already concluded after his trip to the Amazon that the rainforest helps to generate clouds and rain, and that enough rain might fall in the Sahara for abundant growth if only there were trees. Using computer simulations, my colleagues at the Potsdam Institute have shown that a green Sahara would indeed have the tendency to attract humid air from the Atlantic and monsoon rain (Ganopolski *et al.* 1998). In fact, at the beginning of the Holocene, the Sahara was green. Evidence of this is not only obtained from rock art like in the famous "Cave of Swimmers" in the Gilf Kebir (a lake was just outside the cave at the time of the creation of the murals), but also from archaeological findings such as hippopotamus bones. Due to the slow change in the orbit of the Earth around the sun, the conditions for monsoon rain in the Sahara deteriorated over the millennia. But apparently the vegetation did not slowly and gradually decline in step with the orbital change – i.e., in a linear way – but instead was able to sustain itself almost completely up to a certain point in time, after which it suddenly died from drought conditions. Within a relatively short period of time, the Sahara changed into the dry desert it has remained until today. Both data and computer simulations show this sequence of events. The population of the Sahara, which had been scattered over wide stretches of land, had to flee and was crammed together in the Nile valley – a factor which may have contributed to the emergence of the sophisticated culture of the Pharaones.

Will the Gulf Stream Cease to Exist?

Another example of non-linearity in the climate system, which is now rather well-understood, is the Atlantic ocean circulation (Rahmstorf 1997, 1999). Extending along the entire Atlantic, from the Cape of Good Hope in the South up to Svalbard in the North, a gigantic overturning motion of water takes place, in which the surface waters flow northward, sink down and return south at a depth of two to three thousand meters. The Gulf Stream off the North-American coast and its extension to Europe, the North Atlantic Current, are elements in this larger circulation system, which is driven by differences in the density of the seawater. Where the highest densities are reached, in the Greenland sea and in the Labrador sea, the water sinks down, drawing further water northward like the plug-hole of a bath. The amount of water turned over in this way equals twenty times the flow of all rivers of the Earth combined. As the northward-flowing surface water is considerably warmer than the deep water returning south (which after all comes from the arctic), this system functions like central heating for Europe. Huge amounts of heat, corresponding to the output of half a million large power plants, are transported to the North Atlantic. There the heat is transferred to the air and then carried to Europe by westerly winds. This is why it is cold in winter in Europe if the wind comes from the East, but mild if the wind comes from the West. Looking at climate maps one notices that in Northern Germany the climate is roughly five degrees too warm for the latitude – the respective latitudes of Canada are considerably colder, even on the Pacific Coast. In contrast to the Atlantic, the Pacific has no built-in central heating.

Unfortunately, the Atlantic ocean circulation has a drawback: it is not entirely stable. During the past ten thousand years, our heating system has worked without problems, albeit with some minor variations. But in the more distant past, it has stuttered badly and, during the extreme cold events in the glacial period mentioned above, it even repeatedly collapsed completely. This is revealed by the deep sea sediments (Bond *et al.* 1993).

The reason for the strange behavior of the currents has been analyzed with the help of computer simulations and can be easily understood. In order to be heavy enough to sink down, the water of the North Atlantic has to contain enough salt, as salt increases the density of the water. An opposing effect is precipitation, which lowers the density of the water. However, this effect is offset as long as new, salty water continues to flow in from the South. In short, the current flows because the water is salty, and the water is salty because the current is flowing. This is a clear case of a self-sustaining system. If in the computer simulation the precipitation over the North Atlantic is further increased, the current is, at first, only moderately weakened. But at some stage a point is reached when the influx of new salty water becomes too weak. The precipitation dilutes the water, the current becomes even weaker – a vicious circle which eventually leads to the complete breakdown of the current. Here we have a classic example of a non-linear system which regulates itself within certain limits, but which then all of a sudden experiences a dramatic change if these limits are exceeded.

This finding is of concern since due to the anthropogenic greenhouse effect, precipitation in the North Atlantic region is expected to increase. In his article "Unpleasant Surprises in the Greenhouse" (Broecker 1987), the American climatologist Wallace Broecker warned in 1987 that due to the greenhouse effect, man might cause the Atlantic circulation to stop. Since then, several groups of researchers have been working on a closer examination of the stability of this current.

Present knowledge gives no reason for panic, but also no reason for "all clear". It is likely that during the next decades the North Atlantic Current will weaken noticeably – simulations of the various institutes agree on this point. Although this will gradually wind down Europe's central heating, it will not lead to dropping temperatures – for at the same time, over the whole planet, temperatures are rising due to the greenhouse effect, and this effect also prevails in Europe.

Risky Greenhouse

What will happen in the longer term is far less certain. If man continues to enrich the atmosphere of our planet with greenhouse gases such as carbon dioxide, the global climate could get close to the point where the North Atlantic current is interrupted. Due to the inaccuracy of present climate models, it is not possible to date to make a clear statement as to where exactly this point lies and whether and when it might be exceeded.

It is another characteristic of non-linear systems that the prediction of future trends is considerably more difficult than with linear systems. The existence of critical thresholds which, if exceeded, would mean a sudden qualitative change of climate is in principle easily understood. But where exactly these points lie, and the behavior of the climate system near such points, is difficult to calculate. The new findings mean, above all, more uncertainty. The more we drive climate away from its present equilibrium, the higher the risk of unpleasant surprises becomes. Broecker once put it like this: Climate is an unpredictable wild beast, and we are poking at it with sticks.

What could an interruption of the North Atlantic current look like? At the Potsdam Institute we have calculated several scenarios (Rahmstorf & Ganopolski 1999) and found that temperatures in Europe would at first rise significantly, although perhaps slightly less than the global mean temperature – in our scenario by about three degrees until the year 2100. This warming could be followed by an ocean circulation collapse and a sudden drop in temperature back to the pre-industrial level. Over the following centuries, when the carbon dioxide content of the atmosphere will slowly drop again (as mankind is unlikely to infinitely use fossil fuels to the extent we are using them now), temperatures in Europe will fall further and further, until they will finally be about five degrees colder than at present. A roller coaster ride for us Europeans, which would require constant adjustments, first to a warmer and then to a colder climate. At the end the climate would be so cold and dry that agriculture would be hardly possible.

The climatic state that would then be reached – an Earth without the "central heating system" provided by the North Atlantic current – would, according to our calculations, be stable over future millennia. This means that mankind could, by a temporary perturbation of the climate lasting only a few centuries, shift the planet into a completely different climatic state due to its excessive use of fossil fuels.

Whether we should risk such a far-reaching and dangerous interference with the climate system is a political and ethical question. From the purely factual viewpoint of a climatologist, who examines the fluctuations of global climate over millions of years, there is of course no climatic state that is 'better' than another one, and constant change is the rule. There is however no doubt that extensive and rapid climatic change bears substantial dangers for mankind. Our eco-systems, agriculture, and settlement structures are based on current climatic conditions, and every large change can lead to great misery. In decades to come, man will become the dominant cause of climate change. The recognition that the climate has a tendency towards sudden jolts should make us cautious.

References

Blunier, T. *et al.* (1998), Asynchrony of Antarctic and Greenland climate change during the last glacial period. *Nature* **394**, 739-743.

Bond, G. *et al.* (1993), Correlations between climate records from North Atlantic sediments and Greenland ice. *Nature* **365**, 143-147.

Broecker, W. (1987), Unpleasant surprises in the greenhouse? *Nature* **328**, 123.

Ganopolski, A., Kubatzki, C., Claussen, M., Brovkin, V. & Petoukhov, V. (1998), The role of vegetation-atmosphere-ocean interaction for the climate system during the mid-Holocene. *Science* **280**, 1916-1919.

Rahmstorf, S. (1997), Risk of sea-change in the Atlantic. *Nature* **388**, 825-826.

Rahmstorf, S. (1999), Shifting seas in the greenhouse? *Nature* **399**, 523-524.

Rahmstorf, S. & Ganopolski, A. (1999), Long-term global warming scenarios computed with an efficient coupled climate model. *Climatic Change* **43**, 353-367.

Possibilities and Limitations for Flexible Compliance with the Kyoto Targets

Hans-Jürgen Ewers

The German Council of Environmental Advisors (SRU), P.O. Box 55 28, D-65180 Wiesbaden, Germany.

Economic Instruments in Climate Policy: The Kyoto Protocol

Central results of the Kyoto Protocol are the agreement on a quantified emission limitation of global greenhouse gases (-5.2% in the years 2008 to 2012 compared to 1990), the allocation of the reduction target to the industrialized countries which commit themselves to comply with an emission budget (Annex I States), as well as the admission of flexible market-economy instruments for reaching the target.

With the admission of flexible instruments, the Annex I States can choose whether they want to fulfill their commitments by own emission reduction efforts, or whether they want to buy the emission rights from other countries which may not need these rights due to their stage of economic development, or because they have over-fulfilled their commitment thanks to the use of cost-effective emission reduction technology. The accusation that this system was "modern trading with letters of indulgence" is not justified, at least from an economic point of view. Rather, it has the advantage that emissions can be avoided where it is most cost effective to do so. Countries with high reduction costs can buy emission rights, while countries with low reduction costs are able to profit from selling their rights. The full cost reduction potential is used if other parties (industry, associations, individuals) apart from governments can be involved in the trading of emission rights; sector-related solutions alone are inefficient from an economic point of view.

In comparison to a "tax solution", tradable emission rights have the advantage that they take effect where it is ecologically necessary while governmental interference is kept to a minimum. Compliance with the international climate protection targets is ensured by the number of certificates issued. With a tax solution, changes in the surrounding economic conditions (for example the inflation rate, technological progress, economic growth, exchange rate fluctuations) make an ongoing adjustment of taxation rates necessary in order to avoid endangering the environmental objectives. Instead, if there was a market for emission rights all necessary adjustments would be achieved merely by the price-induced adjustments of supply and demand until a balance is reached. In this respect, in the long run a global system of internationally tradable emission rights is suited best to ensure the achievement of the international climate protection policy goals.

The Kyoto Protocol departs from three flexible market-economy instruments (emission trading, joint implementation and clean development mechanism), which are all ultimately based on the fundamental idea of trading pollution rights and which will be shortly explained in the following.

Emissions Trading (ET)

In Annex B of the Kyoto Protocol, the industrialized nations (Annex I States) undertake to limit, or respectively reduce emissions of the greenhouse gases Carbon dioxide (CO_2), Methane (CH_4), Nitrous oxide (N_2O) Hydrofluorocarbons (HFCs), Perfluorocarbons (PFCs) and Sulphur hexafluoride (SF_6) from 2008 to 2012 in comparison to 1990. They can ensure their compliance with the agreed emission budgets either by avoiding emissions themselves or by participating in emissions trading (Article 17 of the Kyoto Protocol). Those entitled to participate in this trade are all states which have committed themselves to a quantitative reduction. The relevant principles, conditions, rules and guidelines for the trade of emission rights, particularly for the monitoring, reporting and verification system, are yet to be worked out.

States which do not fulfill their commitment between 2008 and 2012 by own measures can purchase emission rights not needed by other countries in the international emission permits market. The entire reduction target of minus 5.2 % compared to 1990 will still be achieved, as the increase in the emission budget ("assigned amount") of one Annex I State by the transfer of emission rights is offset by a reduction of the same amount of the emission budget of another Annex I State.

Evidence of compliance with the emission budget in the Annex I States is achieved by the setting-up of greenhouse gas emission inventories (Article 7 of the Kyoto Protocol). These inventories, which have to be submitted to the Annex I States on an annual basis, must consider anthropogenic emissions by sources as well as removals by sinks. Criteria for credits from sinks are currently being worked upon by the Intergovernmental Panel on Climate Change (IPCC).

Joint Implementation (JI)

In addition to the instrument of emissions trading, Article 6 of the Kyoto Protocol provides the transfer of emission reduction units resulting from joint projects between Annex I States (joint implementation). This refers to projects which involve a reduction in greenhouse gas emissions by sources, or an enhancement of removals by sinks.

The transfer of emission rights from JI-projects causes a charging of the emission budget of the recipient country, and an emission credit for the investing country (cf. Schwarze & Zapfel 1998). Emission rights created by JI-projects but which are not needed can also be sold to other Annex I States. The common upper emission limit of all 38 Annex I States, amounting to 94.8 % of greenhouse gas emissions in 1990, is not violated by this instrument of JI. The possibility of crediting the investing country for emission reductions is even likely to stimulate a technology transfer between Annex I States.

Clean Development Mechanism (CDM)

Finally, the instrument of the clean development mechanism enables the Annex I States to acquire emission credits through projects in non-Annex I States if these projects mean a reduction of greenhouse gas emissions in the recipient country compared to the "business-as-usual" scenario (Article 12 of the Kyoto Protocol). Certified emission reductions achieved by CDM-projects from the year 2000 onwards can be used to assist in achieving compliance in the first budget period (2008-2012). It is however still unclear whether emission credits acquired through CDM projects should be tradable between Annex I States or not.

CDM promotes the use of cheap abatement opportunities in countries which have not entered a commitment in the frame of the Kyoto Protocol themselves. The possibility to acquire emission credits through CDM-projects from the year 2000 onwards creates incentives for early activities in climate protection. non-Annex I States can profit from technology transfer. The instrument is problematic insofar as yet there are no clear criteria for determining which projects lead to emission reductions that would not have been achieved without the CDM. However, the lack of criteria for the identification of the so-called "baseline" creates the danger that the Annex I States fulfill their commitment by projects that result in no reduction of emissions compared to the "business-as-usual" scenario, which would undermine the goal of a limitation in global emissions.

"Hot Air" and Other Problems: Opportunities and Limitations in the Realization of the Kyoto Protocol

"Hot Air"

Despite their obvious advantages, the results of Kyoto are not undisputed. The criticism from Germany and some other Parties to the Protocol is directed above all to the fact that the Protocol allows Russia and the Ukraine to keep their 1990 emission levels, although emissions in these countries have since diminished considerably due to the economic breakdown (so-called "hot air" problem). Now it is feared that other countries might buy these emission rights and thus fail to sufficiently fulfill their own commitments agreed in the Protocol. For this reason, a number of states, among them Germany, want to restrict international trade of emission rights.

From an ecological point of view, emission reduction targets that are stricter than those agreed in Kyoto would be desirable and also achievable at low cost if the flexible instruments provided for in the Protocol were used. However, an additional limitation of trade with "hot air" is likely to be politically difficult. In fact, such a "cap" solution could endanger the ratification of the Protocol by a number of Parties and thus also the entire process towards an internationally concerted climate protection policy. Instead of losing the opportunity to establish a global trading system which no country could leave without losing a loss of reputation, one should try to assure that further reduction targets for future periods, to be agreed upon in a targeted new agreement, meet ecological requirements better than the actual commitments do. This is underlined by the demand of the German Enquete Commission "Protection of the Global Atmosphere" to reduce global CO_2 emissions by 50 % until the year 2050 in comparison to 1987 – and this is with a growing global population.

Sanctions

Prerequisites for a functioning trade with emission rights are suitable sanctioning mechanisms. Compliance with the emission budgets laid down in the Protocol has to be monitored and their non-compliance sanctioned. For Annex I States participating in emissions trading, incentives for compliance with their commitment could be created by devaluating sold emission rights if the emission budget is exceeded. The risk of non-compliance lies thus with the buyer (buyer-beware-principle). States willing to sell their rights would therefore be forced to prove how they will keep their commitment at the end of the budget period. Another approach would be that the Member States in the case of non-compliance with other commitments of the Protocol (for example setting up of an emission inventory) have to expect exclusion from trading. If states not involved in international emissions trading face a penalty in case of non-compliance which is higher than the price for emission rights, they will store a sufficient number of emission rights.

Other Rules for the Use of Flexible Instruments in the Kyoto Protocol

To make use of the market-economy instruments laid down in the Kyoto Protocol, further rules are necessary which should consider above all the following aspects:

- Following securities trading, international rules have to be set up which allow a transparent and discrimination-free trade with emission rights.
- For the crediting of emission reductions by CDM-projects, clear criteria have to be developed.
- Finally, guidelines for the crediting of biological sources and sinks have to be worked out.

Compatibility of Emissions Trading with an Ecological Tax Reform on a National Level

The establishment of a system of tradable emission rights is not to be understood necessarily as abandonment of an ecological tax reform, the first stage of which came into force in Germany on April 1, 1999. Rather, the political correction of energy prices represents a long overdue step.[1] Whether this will happen by way of a tax or a permit solution, is of secondary importance. Rather, a correction of energy prices by way of taxation is in principle suitable for creating incentives for a reduction of energy prices on a national level until an international permit system has been established. It is also conceivable that both instruments coexist for a while. E.g., emission sources could choose between a permit or a tax scheme. Also, the obligation to hold permits could at first only apply to a certain group of emitters (such as power plants), while others (for example road traffic) have to pay taxes. The basis for calculating tax rates would be the permit price. Sector-specific differences in price for the same charge or consumption of ecological resources should be avoided for reasons of efficiency.

---

1 For criticism of the draft of a law to enter an ecological tax reform cf. German Council of Environmental Advisors (1999).

Outlook

The flexible market-economy instruments laid down in the Kyoto Protocol show a number of ecological and economic advantages. In this context, the issue of trading with "hot air" represents a distributional problem which should not obstruct the introduction of a system of tradable emission rights. As the implementation of appropriate market-economy instruments will allow the achievement of climate-policy targets at minimum cost, there is reason for hope that the Member States will enter higher reduction commitments once a permit system has been introduced and unfolds its full efficiency.

Apart from the agreement on even more ecologically beneficial emission reduction targets in later periods, future political efforts in climate protection should include above all the following points:

- the introduction of an effective sanctioning mechanism, punishing those exceeding national emission budgets,
- the creation of rules allowing a transparent, discrimination-free trade with emission rights,
- the provision of clear criteria as a measure for determining whether reductions in emissions ensuing from CDM projects should be credited,
- the development of guidelines for accounting of biological sources and sinks and
- the better inclusion of non-Annex I States into climate policy measures. Possible starting points for a cooperation are population growth, the world trading rules as well as the protection of tropical forest regions, which are increasingly endangered by fire clearing.

References

German Council of Environmental Advisors (1999), Position on the draft of a law to enter an ecological tax reform. Public hearing of the Finance Commission of the German Bundestag on January 18, 1999.

Schwarze, R. & Zapfel, P. (1998): *Klimaschutzzertifikate ante Portas: Eine Analyse der Instrumente der internationalen Klimaschutzpolitik nach Kyoto.* Technische Universität Berlin, Wirtschaftswissenschaftliche Dokumentation, Diskussionspapier 1998/18.

Shaping Greenhouse Gas Abatement Strategies – Policy Issues and Quantitative Insights

Christoph Böhringer [1]

Centre for European Economic Research (ZEW), Head of Department of Environmental and Resource Economics, Environmental Management, P.O. Box 103443, D-68161 Mannheim, Germany, e-mail: boehringer@zew.de.

Introduction

Meanwhile, it is commonly agreed upon that climate protection requires a significant contraction of anthropogenic carbon dioxide (CO_2) emissions and other trace gases. Stringent emission constraints on production and consumption activities are likely to produce non-negligible adjustment costs. Given some exogenous global emission reduction profile (which might be delivered from natural sciences) international climate policy focuses on two major issues:

- What is the magnitude, and
- what is the distribution of costs

for alternative greenhouse gas (GHG) abatement strategies to achieve the global reduction target?

In the following presentation I want to illustrate how quantitative economic analysis can contribute to these issues. It will become clear that flexible market instruments, particularly, international emissions trading, play a key role in reducing total economic costs and alleviating potential burden sharing problems.

---

1 Research support has been provided by the Ministry for the Environment and Transport of the German State of Baden-Württemberg. The analysis of "Contraction and Convergence" builds on joint work with Heinz Welsch (University of Oldenburg).The author remains responsible for the views expressed here.

To motivate the role of applied economic analysis, I will first sketch the nature of problems that climate policy faces from an economic point of view. I will then present quantitative assessments of two abatement scenarios and highlight the implications of emissions trading as a flexible instrument on the magnitude and distribution of abatement costs. The first scenario, "Kyoto", actually reflects the implementation of the Kyoto Protocol. The second scenario, "Contraction and Convergence", is much more challenging and investigates the economic consequences of a climate policy in which global emissions are substantially reduced as compared to current levels, and emission rights are allocated in the long run on a per capita basis across all world regions. The quantitative results for both policy scenarios are based on simulations with PACE, a large-scale modeling system designed to analyze the economic implications of environmental policies (Böhringer 1999a). Finally, I will summarize and conclude.

Policy Issues and the Role of Applied Economic Analysis

Buying Global Warming Insurance – Total Costs of Abatement

There is an international consensus that policy measures are required to stabilize greenhouse gas (GHG) concentrations in the atmosphere in order to mitigate man-made climate change. Little agreement exists, however, on the critical level of concentration and the risks of climate change. Consequently, the scope and the timing of emission reduction are highly disputed issues among negotiators of emission abatement strategies. Some parties point to the uncertainty of the risks of climate change. In face of future uncertain benefits they argue against stringent short-term emission reduction and would rather vote for "wait and see" or "learn then act". On the opposite side of the spectrum, other parties go for substantial short-term reduction because they expect large adverse impacts from an increase in global temperature if not immediately acted upon. They also fear that nothing will be done if everyone delays, and consider short-term action necessary to get the political process moving. The negotiation of the Kyoto Protocol is an example of these different perceptions even among the developed countries. While the US, for example, considers the Kyoto targets as quite demanding, other negotiating parties such as the EU would have liked to agree on much higher quantified emissions limitation and reduction objectives (QELROs). To some extent the climate policy debate boils down to varying degrees of risk aversion and the associated willingness of current generations to pay for climate change insurance of future generations.

Conclusion: Climate policy involves trading off the benefits of mitigating global warming with the costs of emission abatement. Despite of major uncertainties, quantitative cost-benefit analysis is therefore a prerequisite for rational policy making.

Alternative Policy Instruments – The Issue of Cost Efficiency

Having agreed on some global emission reduction trajectory, the question arises as to which policy instruments one should choose. Common sense suggests to pick the least-cost instrument that assures the given environmental target. Basic economic theory, then, tells us that cost-effectiveness requires equalization of marginal abatement costs across different sources. In intuitive terms, this means that we can save costs as long as we have not fully exploited the opportunities of shifting from higher cost abatement options to lower cost abatement options.

Figure 1 visualizes the economic efficiency argument for equalizing marginal abatement costs. Marginal abatement cost MAC_1 and MAC_2 for two different countries are shown on the vertical axes while the horizontal axis indicates the emission reduction. Suppose that we have to cut back total emissions by ΔE. One way of doing this is simply to tell each country to reduce its emissions by $\Delta E/2$ (standard setting). The two countries face different marginal abatement costs and there is scope for efficiency gains or total costs savings when country 2 with higher marginal cost abates less and country 1 with lower abates more. The cost-efficient solution is achieved when marginal abatement costs are equalized. The shaded area ABC indicates the efficiency gains as compared to the initial situation where marginal abatement costs differ.

Equalization of marginal abatement costs comes down to levying a uniform emission tax or implementing a market for tradable permit rights. Both instruments assure a cost-efficient solution. In the following, we will focus on tradable permits as a flexible instrument. Note that an emission tax is in fact equivalent to the price of the permit right when it induces an emission level equal to the quantity of tradable permits. The major difference is that taxes are price-based (i.e. the price of emissions are set exogenously) with the emission reduction being determined endogenously while permits are quantity-based (i.e. the emission quantity is set exogenously) with the emission price being determined endogenously.

Our little excursus in basic environmental economics has hopefully clarified the notion of "Flexibility for Efficiency" as adopted in the conference title. In other words, instruments for climate policy should be flexible enough to allow for equalization of marginal abatement costs, i.e. cost-efficiency. When dealing with problems that have – beyond the spatial dimension – an explicit time dimension, cost effectiveness may not only call for equalization of marginal abatement costs across sources (when-flexibility) but over time (time-flexibility). This is actually the case for global warming because the effect of GHG emissions on global warming depends on the accumulated trace gas concentration in the atmosphere. Given a critical concentration level, the concrete time-path of emission reductions over the next few decades is rather unimportant from a natural science point of view, as long as the timepath complies to an overall cumulative emission budget.

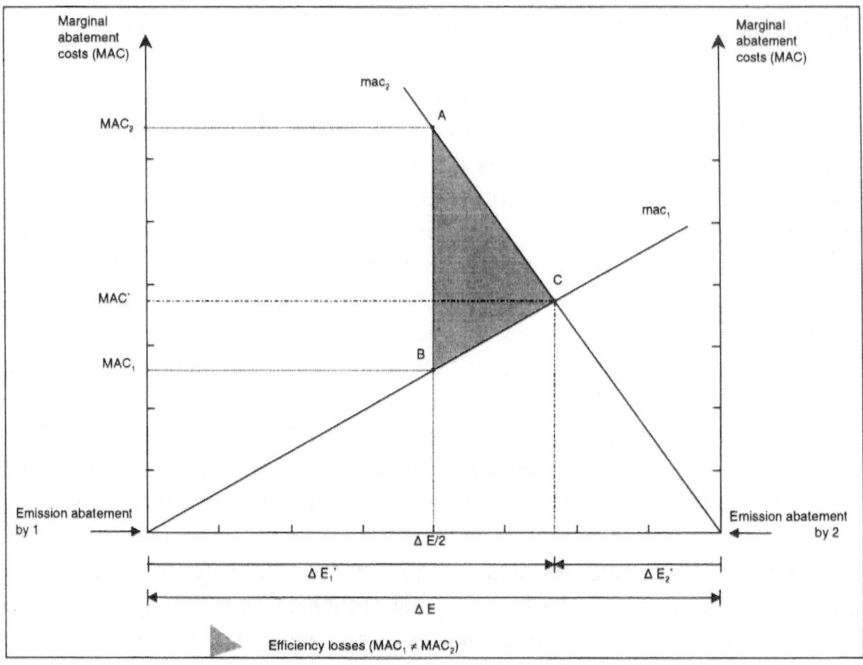

Figure 1: Efficiency Gains from Equalized Marginal Abatement Costs

Cost effectiveness of climate protection strategies may be also affected by the way in which potential revenues from emission permits or emission taxes get recycled into the economy. There is a large theoretical and empirical literature on the question if the use of additional revenues to cut down other existing tax distortions may actually overcompensate the economic costs of abatement, i.e. simultaneously improve environmental quality and economic performance. We do not want to enter the debate on this so-called double-dividend hypothesis in this paper. However, we acknowledge that deliberate revenue recycling may significantly reduce the economic burden of emission abatement.

Conclusion: Given the long-term rather uncertain benefits of climate protections and the more certain, rather near-term costs of abatement, emission reduction strategies will be more acceptable the cheaper, i.e. the more cost-effective, they are. We therefore need systematic and consistent quantitative analysis of how alternative strategies which achieve the same environmental target differ in costs.

Distribution of Costs – The Issue of Burden Sharing

Global warming is a global problem that requires a global solution in the sense that all major emitting parties must be part of some effective abatement strategy. Agreement to future stringent QELROs is only likely when the international parties perceive the distribution of associated costs as fair. For a given global reduction target, the distribution of costs is to a large extent determined by the allocation of emission rights across countries. The debate on equitable sharing of the EU aggregate Kyoto target across EU member countries has already provided an illustrative example of the difficulties involved in identifying an "equitable" burden sharing scheme. The question of a "fair" distribution of abatement costs across countries inevitably involves ethically founded equity criteria. While positive economic theory has nothing to say on equity, it can at least clarify the economic consequences of alternative equity rules.

Conclusion: The prospects of achieving further international agreement on climate protection policies will crucially depend on whether the agreement is perceived to be fair. Quantitative estimates on the economic implications associated with alternative equity rules are a prerequisite for any equity debate.

The Role of Quantitative Economic Analysis

Rational climate policy making requires quantitative estimates on both the total costs as well as the cost distribution of alternative emission abatement strategies. As it is not possible to simulate the economic effects of alternative abatement strategies within the real world modeling economic adjustment is an important tool for gaining policy relevant insights. The use of economic models allows for the systematic and consistent analysis of alternative policy scenarios. Given the fact that the aggregate economic impact is determined by a number of partial effects which may work in opposite directions, it is not wise to rely on basic economic intuition for sound decision making. However, usefulness of applied modeling requires a check of the underlying assumptions. Careful sensitivity analysis to assumptions is a prerequisite for gaining robust insights. While this is a major challenge, it is, at the same time, the major strength of systematic model-based analysis (Böhringer 1999b).

Climate Protection Policies and the Implications of International Emissions Trading

The Economic Impacts of Implementing Kyoto

The Kyoto Protocol constitutes a milestone in climate policy as major emitting regions have accepted – for the first time – legally binding emissions targets. In concrete, developed countries – the so-called Annex B parties – have committed themselves to reducing greenhouse gas emissions on average by 5.2 % from 1990 levels in the budget period of 2008 to 2012. Table 1 indicates the Kyoto targets which differ across signatory parties. Developing countries have not adopted any emission constraint under the Kyoto Protocol.

Table 1: Quantified Emissions Limits under the Kyoto Protocol (UNFCCC 1997)

Region	*Commitments (Percentage of 1990 Base Year Greenhouse Gas Emissions)*
USA / Canada	93.1
European Union (EU15)	92.0
Japan (JPN), Australia (AUS), New Zealand (NZL)	96.8
Economies in Transition (EIT)	98.3

In our quantitative analysis, we distinguish two cases which reflect different assumptions on the flexibility of abatement policies within the Annex B country group. Under *NTR* (emission permits are not traded internationally) Annex B countries can trade emission rights as allocated under the Kyoto Protocol only within domestic borders – there is no international trade in permit rights. Under *TRD* (trade in permit rights across all world regions) Annex B countries can buy (sell) emission rights from (to) another Annex B country.

In our dynamic analysis of these policy scenarios, we assume for the post-Kyoto time horizon that Annex B parties maintain the Kyoto targets from 2012 onwards, whereas no emission constraints apply to the developing countries.

Figure 2 illustrates the potential losses in lifetime consumption for representative agents in aggregate world regions which are central to the greenhouse gas issue. Losses are reported in percentage change as compared to a business-as-usual policy (*BaU*) without any GHG emission constraint.

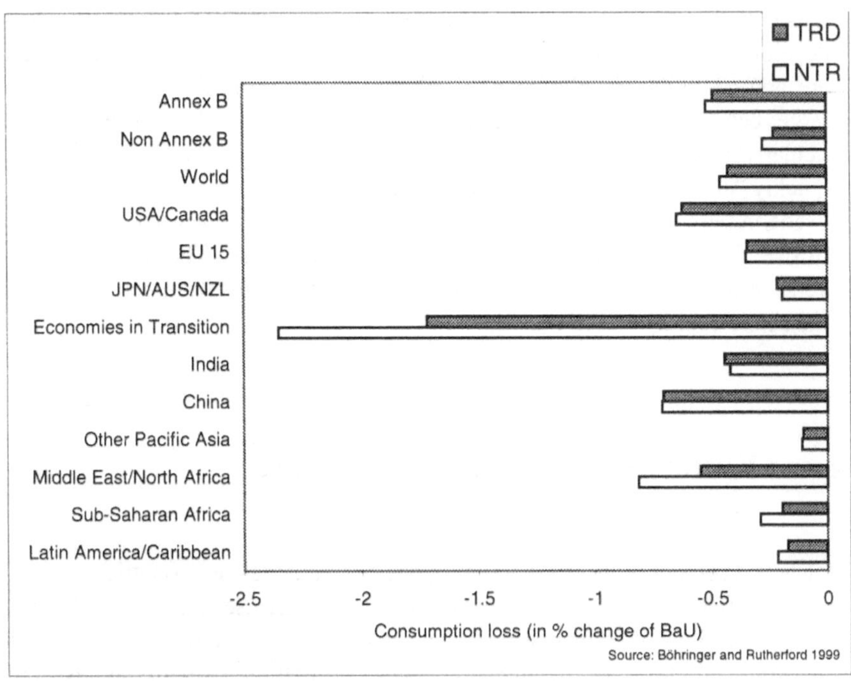

Figure 2: Impacts of Kyoto on Lifetime Consumption (% change as compared to *BaU*)

The key insights from the numerical analysis can be summarized as follows (see Böhringer & Rutherford 1999):

- Implementation of Kyoto not only involves significant adjustment costs for developed Annex B countries, but also for the developing non-Annex B countries. While economic adjustment costs, due to stringent emission constraints in the developed world, are straightforward, the negative impacts on developing countries require further explanation. The reason is that Annex B countries pass on part of their adjustment costs through higher export prices to trading partners in the developing world. non-Annex B countries suffer from an "income effect" due to reduced export demand from the constrained developed world. On the other hand, they may gain from a "substitution effect" due to increased competitiveness in producing energy-intensive goods. A third international spill-over effect stems from the drop in international fuel prices because of the reduced demand in fossil fuels by Annex B countries. This is beneficial for fossil fuel importers and harmful for fossil fuel exporters. Our quantitative results indicate that the aggregate outcome of all three effects, which often is referred to as terms of trade effect, is negative for the developing countries.
- Permit trade between Annex B countries reduces total costs of Kyoto both for the Annex B country group as well as for the non-Annex B country group. The

reason is that one can exploit cost differentials in emission abatement across developed countries. While the aggregate efficiency gains of *TRD* versus *NTR* are non-negligible (in the order of 10 % of *NTR* costs) they are not overly large, because there are only limited low cost options within Annex B (particularly those delivered from the economies in transition, EIT) which are quickly "absorbed" by large polluters with high abatement costs.
- Trade in permits may not be beneficial for all countries when the distribution of overall efficiency gains are distributed via the market. This is because adverse terms of trade effects can dominate efficiency gains from permit trade at the single country level.

Beyond Kyoto: Contraction and Convergence

The Kyoto Protocol reduces GHG emissions by industrialized countries as compared to 1990 levels. However, global emission will still rise significantly for reasonable projections of economic growth and fossil fuel consumption in the developing world. Stabilization of greenhouse gas concentrations in the atmosphere, hence, will not only require further emission reduction by Annex B countries, but also restrictive emission constraints for developing countries.

Inclusion of the developing world in some long-term cooperative agreement immediately raises the critical issue of burden sharing. The developing countries have so far refused any abatement commitment, mainly because they fear negative effects of emissions limitation on their economic development. Also, before committing themselves to reduction targets, they demand primary action by the developed world with large historical emissions.

One key proposal for greenhouse gas abatement which entails emission constraints for the developing countries places emphasis on a long-term equitable per capita distribution of emission rights (egalitarian approach) while imposing a significant *contraction* of anthropogenic GHG emissions over the next 50 years. This proposal is unlikely to be acceptable for industrialized countries with currently high per capita emissions, unless the transition path allows for long-term "smooth" adjustment towards the terminal point. Figure 3 presents the per capita trajectory for a politically feasible *Contraction and Convergence* profile where global carbon emissions are cut back by 25%, compared to 1990 emissions and per capita emissions converge linearly from current levels to equal per capita entitlement in 2050 (see Böhringer & Welsch 1999).

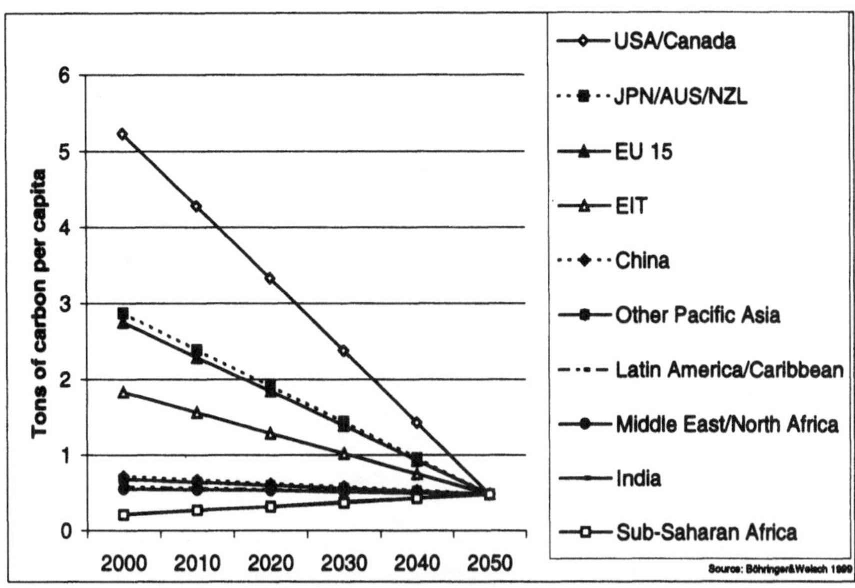

Figure 3: Per Capita Emission Rights under Contraction and Convergence (in tons of carbon per capita)

Figure 4 summarizes the economic impacts of implementing this *Contraction and Convergence* timepath for a no trade case *NTR* and a full trade case *TRD*. As previously done, we measure economic implications with respect to a business-as-usual scenario *BaU* where no emission constraints apply.

The quantitative results deliver the following major insights (see Böhringer & Welsch 1999):

- Emissions trading reduces the economic costs of contracting global emissions by more than half compared to the *NTR* case. This magnitude of efficiency gains from trade reflects the tremendous range in marginal abatement costs across the different regions ranging from zero to 1500 $US per ton of CO_2 (in 2050).
- Trade in permits improves the economic well-being of all regions compared to the *NTR* case. Hence, where-flexibility constitutes a no-regret strategy for implementing *Contraction and Convergence* for all countries.
- Major opponents to emissions restrictions from the developing world such as Africa, Latin America, the Middle East or India are able to improve their economic welfare even over business-as-usual (*BaU*).

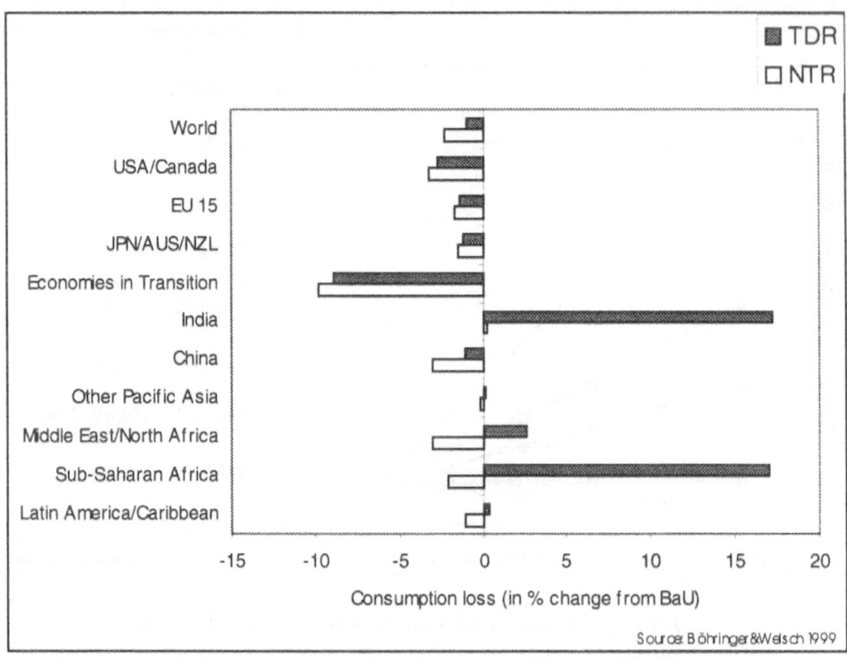

Figure 4: Impacts of Contraction and Convergence on Lifetime Consumption (% change as compared to *BaU*)

Summary and Conclusions

The prospects of achieving sustainable climate protection crucially depend on the magnitude and distribution of economic costs associated with emission abatement strategies. Flexible instruments such as international emissions trading allow the exploitation of differentials in abatement costs across regions and therefore promote cost-efficient solutions. Our quantitative analysis emphasizes the point that where-flexibility provides more cost savings the more stringent the global reduction target becomes. In order to buy the same level of climate change insurance we spend significantly less money under a tradable permit system than without emissions trading. To put it differently, for a given amount of money, flexibility allows the purchase of a higher level of abatement, i.e. risk insurance.

Emissions trading between the developed and the developing countries may considerably relax the problem of equitable burden sharing. In fact, without where-flexibility it will be difficult to implement long-term abatement strategies which include developing countries and comply with basic equity criteria.

Positive economic theory generally has little to say on equity issues. The typical device for an economist is to separate efficiency from equity considerations. Pushing for flexibility, the economists care for the minimization of the total *cost-pie* and then refer to (non-distortionary) lump-sum transfers in order to meet some exogenous equity criterion. However, in political practice it is often not possible to separate efficiency from equity. There are, typically, barriers to a first-best efficient solution due to the unavailability of lump-sum instruments. In this context, applied economic analysis can at least support rational decision making by quantifying the potential trade-off between efficiency and equity for alternative policy strategies.

References

Böhringer, C. (1999a), *PACE – Policy Assessment based on Computable Equilibrium, ein flexibles Modellsystem zur gesamtwirtschaftlichen Analyse von wirtschaftspolitischen Maßnahmen*. ZEW Dokumentation, Mannheim.

Böhringer, C. (1999b), Die Kosten des Klimaschutzes. *Journal of Environmental Law and Policy* (ZfU) **3**, 369-384.

Böhringer, C. & T.F. Rutherford (1999), *World Economic Impacts of the Kyoto Protocol*. In Welfens, P.J.J., Hillebrand, R. & Ulph, A. (eds.), Internalization of the Economy, Environmental Problems and New Policy Options. Springer, Heidelberg/New York, forthcoming.

Böhringer, C. & H. Welsch (1999), *C&C – Contraction and Convergence of Carbon Emissions: The Economic Implications of International Emissions Trading*. ZEW Discussion Paper No. 99-13, Mannheim.

UNFCCC (1997), *United Nations Convention on Climate Change, Kyoto Protocol to the United Nations Framework Convention on Climate Change*. FCCC/CP/L.7/Add1, Kyoto.

Design of the Flexibility Mechanisms: An NGO's Perspective

Nathalie Eddy

US Climate Action Network, International Coordinator, 1367 Connecticut Avenue, N.W., Suite 300, Washington, DC 20036, USA, e-mail: cielne@igc.org.

Since the start of the climate negotiations, Climate Action Network (CAN)[1] has played an active and influential role in the development of international climate policy. Following the adoption of the Kyoto Protocol in December 1997, many CAN groups devoted attention to the flexibility mechanisms. Buenos Aires was an opportunity to focus on the critical elements of the mechanisms and to begin ensuring strength in their rules and institutions. The criteria laid out in this presentation reflect the global CAN position that was reached at the fourth Conference of the Parties (COP4) in Buenos Aires last fall.[2]

This paper explores the roles for the mechanisms, establishes guiding principles to ensure their environmental strength, and highlights some of the specific criteria identified by CAN as essential for their environmental effectiveness and economic efficiency. As negotiators and interested stakeholders in the climate change negotiations work towards COP6 decisions, it is crucial that the rules and institutions for the mechanisms be developed in such a way as to guarantee the environmental integrity of the Kyoto Protocol. This environmental integrity is the foundation upon which the economic value of future credit will depend.

---

1 Climate Action Network represents 281 environment, development and energy NGOs from all continents, with more than 10 million members worldwide.

2 The full global COP4 CAN position paper is available to download at www.climatenetwork.org.

Role of the Kyoto Mechanisms

The Kyoto Protocol requires Annex I Parties to achieve their commitments primarily through domestic action. A vast majority of CAN groups worldwide believe that the best way to ensure the necessary emphasis on domestic action is to apply a quantitative cap to the percentage of an Annex I Party's commitment it can achieve abroad.

Domestic action has many advantages, including its ability to induce Annex I countries to modify their long-term emissions trends and to stimulate technological innovation. Changes in long-term emissions trends and technological advances will in turn reduce the cost of the stronger emissions reductions targets Annex I countries will adopt in the future. However, properly designed mechanisms under the Kyoto Protocol offer increased flexibility to Annex I Parties in achieving their commitments, while also potentially reducing costs.

Environmental Integrity and Overarching Principles

The primary purpose of the mechanisms is to help Parties achieve the Convention's objective of "stabilization of greenhouse gas (GHG) concentrations in the atmosphere at a level that would prevent dangerous anthropogenic interference with the climate system." The mechanisms present both an opportunity and a challenge to all Parties interested in addressing the urgent problem of climate change. On the one hand, they present an opportunity to Annex I countries to help achieve their Kyoto commitments through cost effective mechanisms. On the other hand, the mechanisms present a challenge to Parties, business, and non-governmental organizations (NGOs) to ensure that the credits generated by each of the mechanisms fulfill the ultimate objective of the Convention and supports the commitments of the Kyoto Protocol.

Consequently, any rules for the mechanisms that result in undermining the Kyoto Protocol and/or the Convention will lead only to emissions credits with little certainty or economic worth. The guiding principles for the design of the rules should ensure the mechanisms are environmentally beneficial, economically efficient, institutionally independent, transparent, and verifiable. Each of these five criteria must be carefully evaluated and reported upon by independent entities if the mechanisms are to function as viable options for climate change mitigation under the Kyoto Protocol.

Concerns of the Environmental NGO Community – Potential Loopholes of the Mechanisms

Many groups of the environmental non-governmental organization (ENGO) community hold two major concerns regarding possible environmental pitfalls presented by the mechanisms. First, as explained above, the mechanisms represent an option by which major emitters could potentially avoid the need for significant domestic emissions reductions. Second, if the rules are not sufficiently stringent and accurate, the use of the mechanisms could result in an actual global increase of GHG emissions. Targets that accommodate for a level of emissions growth that will never be achieved, or "hot air," also play a significant role in the global increase of emissions.

CAN has elaborated a number of criteria and guiding principles which, if implemented, would help guard against the potential loopholes of the mechanisms. Rules that support the following criteria for each of the mechanisms will increase the opportunities for environmentally viable and economically efficient climate change mitigation.

Elements of a Credible Clean Development Mechanism (CDM)

Article 12 identifies two primary objectives for the CDM:

- to assist non-Annex I Parties in achieving sustainable development and in contributing to the ultimate objectives of the Convention, and
- to assist Annex I Parties in achieving their commitments under the Kyoto Protocol and the Convention.

Article 12 lays out fundamental requirements for the CDM, while also leaving many remaining decisions regarding the rules and institutions up to future elaboration and negotiation of the Parties. Key to determining the credibility of the CDM will be the Parties' compliance with the following requirements of Article 12: voluntary participation; real, measurable and long-term benefits; and environmental additionality. The second factor will hinge on Parties' ability to recognize and elaborate additional criteria that are necessary to complete the objectives of Article 12. CAN has laid out a number of additional criteria, the further elaboration of which will follow a brief overview of Article 12 requirements.

Article 12 Requirements

Voluntary participation – The voluntary participation of all Parties involved in a CDM project must be consistently upheld. However, while "voluntary participation" alone is important, it falls short of ensuring adequate communications between the capital city of a host country and the site where the CDM project will actually be implemented. CAN advocates defining voluntary participation to include the necessary consultation and participation of local communities and indigenous peoples in the host country.
Real, measurable and long-term – In requiring that the CDM generates only real, measurable and long-term reductions, Article 12 ensures that the CDM can not result in global increase of GHGs.
Additionality – Article 12 identifies the need for reductions in emissions that are additional to what would have otherwise occurred in the absence of the CDM project. Financial and technological additionality should also be part of the CDM's additionality tests.

- *Environmental additionality* – Environmental additionality is essential as it ensures the credits generated by a CDM project are above and beyond what would have otherwise occurred. In the absence of such an environmental additionality test, credits could be generated for reductions that represented business as usual. A key factor in the measurement of environmental additionality is the establishment of benchmarks or baselines. These benchmarks or baselines must be chosen to represent the high performance end of practice. Because CDM projects will include Parties that have not taken on commitments, this critical methodological issue of environmental additionality must be adequately addressed. Otherwise, net global emissions could increase.
- *Financial additionality* – The financial additionality requirement checks that investment in CDM is in addition to Parties' obligations to the Convention and to Overseas Development Assistance (ODA) funds. The CDM should under no circumstances channel funds away from other existing financial assistance obligations of Annex I Parties.
- *Technological additionality* – CAN would add technological additionality to the list in additionality requirements of Article 12. Technological additionality would ensure that CDM projects do not replace technology transfer obligations of Annex I under the Convention and the Kyoto Protocol.

In addition to the Article 12 requirements, CAN has identified a number of additional criteria that would further support the environmental integrity of the CDM.

CAN Criteria for the CDM as Identified in COP4 CAN Position Paper

Through the coordination of international drafting groups, CAN groups from all over the world came together to highlight CDM criteria necessary to support the objectives of the Convention and the Kyoto Protocol.

Use of appropriate local technologies – The use of appropriate local technologies, and the introduction of demonstrated cutting-edge technologies must be required practice for all eligible CDM projects. The absence of such appropriate local and demonstrated cutting-edge technologies could potentially allow CDM projects to be implemented to the detriment of the host country. In establishing which types of technologies are appropriate, there are a number of project types that should be categorically excluded from the CDM portfolio. These unacceptable project categories include large hydro, coal and nuclear projects.

Promote development of institutional capacities – A proactive approach to support the development of institutional capacities in host countries needs to be undertaken as soon as possible. In order to ensure that CDM projects support national and local priorities of the host partner, host countries would need to put into place CDM management entities, policies, programs and strategies with clear project selection criteria prior to any consideration of a project being approved for credit. However, in order for this to be possible, Annex I countries need to make a commitment to promote and develop the institutional capacities in host countries.

Sustainable development – While sustainable development is explicitly one of the primary objectives of Article 12, the levels at which this objective should be achieved are not indicated. In assessing the extent to which sustainable development is reached through a CDM project, the evaluation should be performed at both national and local levels. Such a multi-level approach to sustainable development assessment would help to ensure that the host country priorities are being upheld in both the capital and project site.

Strict guidelines and transparency – The CDM's governance structure and mechanisms for certification, verification and review should be transparent to civil society. There are a number of modalities that could contribute to the guarantee of strict guidelines and transparency throughout, including:

- annual CDM reports from Parties
- life-cycle assessment for all CDM projects
- on site spot checks of CDM projects by independent monitoring teams to confirm:
    - achievement of reported certified emission reductions (CERs)
    - promotion of sustainable development
    - transfer of promised benefits to local communities
- incorporation of the full range of stakeholders
- regular review of CDM guidelines

CER serialization – The CERs generated by CDM projects should be separately accounted for, and tracked as to year, origin and specific project. There should be an international registry of CER transactions that is maintained in an up-to-date way and is fully accessible to the public. In addition, any standards that are established to evaluate CER transactions should be transparent.
Certification of the certifiers – In order to ensure the consistent competence of the CER certifiers, the Executive Board (EB) should establish a process to certify the certifiers. This process should be transparent to civil society and should include a conflict of interest test to ensure that the certifiers hold no financial or institutional interests in the CDM project in question.
Equitable flow – To date, the geographic distribution of projects in the activities implemented jointly (AIJ) pilot phase has not been regionally balanced. The varying capacities of potential host countries to actively and informedly take part in the AIJ Pilot Phase contributed to the regional imbalance. The CDM must provide for a regionally equitable flow of projects. This will require greater Annex I support for institutional capacity building in host countries to provide a wider base of potential host countries for the CDM.
Equitable benefits sharing – Annex I Parties should not be the only partners in CDM projects who are able to benefit from the accrual of CERs. Host countries should also be allowed the benefits of CER accrual. Such an arrangement could be made prior to the implementation of a CDM project, and would be on a project-to-project basis.
Consistent with other Conventions – The rules and institutional structure for the CDM must be consistent with the regulations and priorities set out in relevant Convention bodies such as the Biodiversity and Desertification Conventions.

The above CAN criteria, in addition to the Article 12 criteria would help to establish an environmentally sound CDM that supports the Convention and Kyoto Protocol objectives.

Elements of a Credible JI and Emissions Trading Regimes

The joint implementation (JI) guidelines for environmental additionality, monitoring and reporting, and verification and certification should be as rigorous as those for CDM. It is also important to reiterate that large hydro, coal and nuclear projects should not be eligible project categories under JI.
Prerequisites for participation – As JI projects and emissions trading engage only Annex I Parties who have taken on binding commitments under the Kyoto Protocol, there are a number of prerequisites that must be in place in each country before it is permitted to transfer or acquire parts of assigned amounts.

- Ratification of the Kyoto Protocol

- Compliance with requirements of Articles 5 and 7 of the Kyoto Protocol regarding inventories and national communications.
- Ratification of, and compliance with the compliance regime pursuant to Article 18 of the Kyoto Protocol

Adequate and timely reporting – In order to help ensure transparency and allow for interested NGOs or other stakeholders to monitor and/or review the progress of a specific JI project, the rules on tracking and transferring emission reduction units (ERUs), CERs and parts of assigned amounts (PAAs) must provide for real-time public access to information. Similar to the criteria laid out for the CERs of the CDM, distinct serial numbers must identify all ERUs generated by JI and PAAs transferred through emissions trading.

Transparent tracking and transferring of ERUs and PAAs – Transparency and access to information is key to ensuring the credibility of JI and emissions trading regimes. The mechanisms' frameworks must be set up in such a way as to allow interested bodies to monitor the ERUs and PAAs that Parties transfer and acquire.

Buyer-seller liability – A liability regime that is based on a buyer-seller hybrid model would contribute to investor certainty while also providing incentives for credible credits. Such an approach would clearly be dependent on national reporting requirements that provide for adequate and timely information in a transparent manner.

JI and emissions trading proceeds – In order to create a level playing field among the credits generated by the three mechanisms, JI and emissions trading investments should allocate a portion of the proceeds (similar to the portion allocated under the CDM) to adaptation and mitigation activities in the countries most vulnerable to climate change.

Decrease global emissions – Strong, transparent and verifiable additionality tests for JI are key to ensuring that global emissions do not rise due to loose project implementation and an overage in assigned amount export. The less hot air that is permitted to be traded, the greater the decrease in global GHG emissions. The credibility of the rules and institutions of the flexible mechanisms will hinge on the achievement of such a global decrease in GHG emissions.

Common Guidelines

The preceding sections demonstrate that there are a number of common guidelines that are key to the environmental effectiveness and economic efficiency of the Kyoto mechanisms. The principles of transparency, accountability and credibility must be consistently upheld to the highest of standards. Certain project types such as large hydro, nuclear and clean coal should not be eligible for consideration as CDM or JI projects. Also, any decision on the inclusion or exclusion of sinks projects in the CDM should be postponed until the Subsidiary Body for Scientific and

Technological Advice (SBSTA) has considered the Intergovernmental Panel on Climate Change (IPCC) Special Report on land-use, land-use change and forestry (LULUCF). Accessibility of information will greatly contribute to the credibility of all mechanisms-related activities. Serialization of all CERs, ERUs, PAAs, and assigned amounts (AAs) provides the necessary information for a buyer-seller hybrid liability regime. All mechanisms should make equal contributions to adaptation and mitigation support for countries most vulnerable to climate change.

Environmental Integrity and Strong Rules = Valuable Credits

It is in all stakeholders' interest to take the opportunity to ensure that strong rules, regulations and institutions for the mechanisms lead to a decline in global GHG emissions and thus contribute to the environmental integrity and economic value of credits generated by the mechanisms.

Project-Based Instruments: Economic Consequences of the Kyoto and Buenos Aires Framework and Options for Future Development

Axel Michaelowa

Head of Research Program "International Climate Policy", Hamburg Institute for International Economics (HWWA), Neuer Jungfernstieg 21, D-20347 Hamburg, Germany, e-mail: a-michaelowa@hwwa.de.

Abstract

The implementation of activities aimed to mitigate global greenhouse gas emissions is more cost-efficient in developing countries than in most of the industrialized world. Thus the Kyoto Protocol allows industrial countries to finance emission reductions in developing countries through the clean development mechanism (CDM). It also allows joint implementation (JI) between industrialized countries. Both instruments will likely attract billions of dollars per year. Nevertheless, there are differences that will impact on the attractiveness of the instruments and might lead to a lower flow of funds to developing countries than expected. The major differences are the early crediting of CDM reduction while JI is not subject to an adaptation tax. On the basis of the Protocol one cannot decide which instrument is more attractive – that remains open and will depend on the decisions of future rounds of negotiation that might align the two mechanisms. It depends particularly on the decision whether the CDM will allow unilateral, bilateral or only multilateral project implementation. Moreover, the attractiveness of both mechanisms depends on the stringency of baseline methodologies.

Flexibility Mechanisms in the Climate Negotiations

From an economic point of view, it is efficient to give countries with emission targets a maximum of flexibility concerning the location of emission reduction due to the global mixing of greenhouse gas emissions. Thus, the cheapest measures should be taken first regardless where they take place. However, incentives for long-term innovation have to be provided to ensure that short-term savings do not lead to higher long-term costs (Michaelowa & Schmidt 1997) and/or detrimental social-economic effects on the country where they take place.

The issue whether countries have to reach their greenhouse gas emission targets by domestic action alone or are allowed to credit emission reduction reached through projects abroad has been a major issue in the international climate negotiations from their beginning. The United Nations Framework Convention on Climate Change (UNFCCC) recognizes the principle of global cost-effectiveness of emission reduction in Art. 3, 3. Thus it opened the way for flexibility. As it did not fix a binding emission target for any country, the need to invest in foreign emission reduction was not pressing. As industrial countries and countries in transition agreed legally binding emission targets in the Kyoto Protocol, they now have to start emission reduction in earnest, and therefore are interested in cost effectiveness and flexibility. Concerning the organization of emission reduction abroad, four distinct possibilities have been allowed by the Kyoto Protocol (UNFCCC 1997) – two on a macro and two on a micro level. Many rules remain unclear but the 4th Conference of the Parties (COP4) in Buenos Aires could not take a decision on clarification. Instead, it elaborated a workplan to fix open issues by 2000.

The first and most far-reaching macro mechanism is an agreement on joint targets or "bubbles" (Art. 4). This is done by the European Union, which has negotiated a joint target and distributed it to the member states. As the developing countries currently do not wish to set targets, this way is only open to industrial countries. The second possibility is emissions trade – but after Kyoto this is also only open to industrial countries (Art. 17). These two mechanisms are covered by several presentations – I will concentrate on the two distinct ways of micro or project-oriented emission reduction credited to the investing country. The first possibility is only applicable between industrial countries and named "joint implementation" (JI) (Art. 6). One has to be careful not to mix it with the general concept of JI discussed already in the negotiations leading to the Rio Conference and ever since. The second possibility is applicable to host countries without emission targets, i.e. developing countries. It shall be coordinated by a so-called "clean development mechanism" (CDM) that has only vaguely been defined (Art. 12). The detailed rules are described below.

The Provisions of the Kyoto Protocol Concerning the Clean Development Mechanism and Joint Implementation

Clean Development Mechanism

Art. 12 of the Kyoto Protocol outlines the CDM. It states in paragraph 3 that investing countries get credit for certified emission reductions from CDM projects provided "benefits" accrue to the host country (Art. 12, 3a). Crediting shall be only allowed until a certain percentage of the emission target is reached (Art. 12, 3b) that remains to be defined. It is unclear whether crediting up to this quota is in full or only partial. Besides countries, companies and other entities are allowed to invest and execute projects (Art. 12, 9).

The CDM shall cover its administrative budget through project revenues. Moreover, a "part" of these revenues shall be used "to assist developing country Parties that are particularly vulnerable to the adverse effects of climate change to meet the costs of adaptation" (Art. 12, 8).

It remains open who does certification of emission reduction but it shall be done by independent bodies (Art. 12, 7). The project criteria remain the same as in the 1995 decision on a pilot phase for JI projects without credits (activities implemented jointly, AIJ) (Art. 12, 5).

Joint Implementation

Article 6, 1 allows industrial countries to acquire emission permits through investment in emission reduction or sequestration projects in other industrial countries. The criteria for projects are the same as in the AIJ pilot phase (Art. 6, 1a and b). Emission permits created in that way are to be considered equal to emission permits from emission trade under Art. 17 (Art. 3, 10 and 11). Emission permits cannot be acquired if annual reporting requirements have not been met or the reports do not comply with the binding rules (Art. 6, 1c). If a review team has doubts about the compliance of the host country the permits shall still be tradable but are "frozen" until the doubts are resolved (Art. 6, 4).

Evaluation of Obvious Differences between CDM and JI

Even if the provisions of the Kyoto Protocol leave much space for interpretation and clarification, some major differences between the CDM and JI can be recognized:

- While CDM credits accrue from 2000, JI credits only accrue from 2008.[1]
- JI credits are subtracted from the host country emission target while CDM hosts have no targets.
- While it is only generally stated that acquisition of JI credits shall be "supplemental" to domestic action acquisition of CDM credits shall only cover a "part" of the emission target. The latter is clearly more stringent if supplementarity is defined in loose, qualitative terms while the "part" is fixed as a percentage. If supplementarity on the other hand is defined as a fixed quota being less than 50% of the emission reduction and "part" is a quota higher that 50% the latter is more loose. For the discussion of different quota types see 4.4. below.
- While JI credits are freely tradable this is not clear for CDM credits – at least many observers state that host countries should not be allowed to trade in CDM credits while secondary trade of credits accruing to the investor countries cannot be prevented.
- While JI credits can accrue from sequestration projects (e.g. afforestation) this is not clear for the CDM.
- CDM projects have to pay an adaptation tax and an administration fee which is not the case for JI projects.
- CDM credits come into being only after certification by an independent body; this is not clear for JI.
- JI credits cannot be acquired if the investor country does not meet the Kyoto Protocol requirements for reporting its national emissions; no such rule exists for CDM credits.
- JI credits will be frozen if a participating country's compliance with the rules of the Kyoto Protocol is in doubt.
- The CDM will have some institutional structure, at least an Executive Board, while there are no compulsory multilateral JI institutions.

The effects of these differences on the attractiveness of the instruments will be summarized in Table 1:

---

1 This is challenged by a number of countries and experts that argue for early JI crediting (see e.g. CCAP 1998c)

Table 1: Impact of Rule Differentials on Relevant Parameters

	Costs per credited ton of emission reduction	**Integrity of the climate regime**	**Trans-action costs**	**Size of positive project externalities**	**Attractiveness of long-term projects**
Early CDM credit	lower	no impact	lower	no impact	higher
No target for CDM host countries	lower	lower	no impact	higher	no impact
Stricter CDM supplementarity rules	higher	no impact	higher	higher	lower
Restricted trade in CDM credits	higher	no impact	higher	no impact	lower
Sequestration credits in JI	lower	lower	no impact	lower	higher
CDM adaptation tax and administration fee	higher	no impact	higher	no impact	no impact
CDM credit certification	higher	higher	higher	higher	higher
JI credit blockade if reporting is not met	higher	higher	no impact	no impact	no impact
Freezing of JI cre-dits if non-compli-ance is suspected	higher	higher	no impact	no impact	lower
Compulsory CDM structure	higher	higher	higher	higher	higher

The table shows that there is no clear-cut disadvantage for any instrument – the rule differences work in both directions. The overall effects of the rules will depend on their exact definition. It can be stated, though, that transaction costs of JI will be certainly lower than those for the CDM.

Possible Development of CDM and JI Taking into Account Unclarified Rules

The following will discuss the spectrum of scenarios that is possible under the Kyoto rules. One major unknown factor that will not be discussed in detail is the development of emissions trading under Art. 17. If an easy system of emissions trading comes along, it is likely to crowd out JI to a big extent as it will be much easier for potential JI host countries to organize the emissions reduction themselves and then trade surplus emission permits. Surely, "hot air" trading with countries in transition would be the cheapest possibility.[2]

Advantages for JI

It is unlikely that JI will be burdened with much regulation as the overall target level for industrial countries would remain unchanged in case of lax baselines. Thus in any case verification and certification costs will be higher for the CDM than for JI. Another general point is that capacity building needs will be higher under the CDM than in the case of JI due to the better human capital in countries in transition.

The CDM could be stifled by prohibitive financing requirements for adaptation projects that raise the costs for investors. The Brazilians proposed (Gylvan Meira Filho 1998:42) that the respective adaptation tax should be set at 3% while business representatives asked for a tax around 1% and environmental non-governmental organizations (NGOs) for at least 10% (Greenpeace 1998). Estimates for administration costs vary widely between 1 and 15%; the Brazilians set the fee at 3%.

Many host countries fear the CDM reaps "low-hanging fruit" that will not be available when they take up commitments in the future. This consequence will only arise, though, concerning projects that can be "stored" until the date where the commitment is taken. This applies to land-use projects such as forestry, but not to investments in infrastructure with fixed lifetimes that expire before the commitment is taken. In any case, host country governments could require project proponents to pay a tax that enters a compensation fund to be disbursed when a commitment is taken. Such a tax would be a disadvantage.

Small country governments would prefer a compulsory CDM institution as it would reduce their transaction costs and make more probable that they get a share of the projects. Small investors have no chance to develop bilateral projects on

2 The term "hot air" refers to emission targets for transition countries that even given a major economic upswing will not attain the historic emissions of 1990 during the commitment period. This will enable them to trade with part of their assigned amounts which are not backed by real emission reductions.

their own. They are interested in an emission credit which is insured against failure and which bears no unexpected transaction costs. Moreover, it should be usable to cover own emission reduction obligations as well as to be transferred. A multilateral fund supervised by an UN organization would fulfill all these criteria and be an ideal solution for small investors.

CDM performance would be measured against the parameters of:

- number of projects approved,
- cost-efficiency,
- "real, measurable, and long-term benefits related to the mitigation of climate change" (UNFCCC 1997, Art 12(5b)).

Thus the CDM will permanently be torn between two extremes:

- Lax approval of as many projects as possible, disregarding verification and control.
- Over-controlling, costly, bureaucratic procedures.

Given the nature of organizations, the second case seems more probable. This would especially apply if the idea of Haites (1998) would be implemented that the Executive Board should take the decision of certification of each project's credits. Crucial is also how the methodology to calculate baselines will be designed. For a discussion of the baseline issue see Michaelowa (1998).

A CDM that does not accept sink projects would lead to a major advantage for JI as the sink potential in CDM countries is much higher.

Advantages for the CDM

A major advantage that more than weighs up the disadvantages is the early crediting of CDM. Moreover, infrastructure built up during the AIJ pilot phase can be reused while it seems unlikely that it will be kept in place in the JI countries where crediting only begins in 2008.

If the CDM was a small, efficient clearinghouse or only a project exchange lowering transaction costs for investors, it would have a distinct advantage compared to JI. Governments of big emitters will favor such a small-scale CDM as transaction costs for the bilateral approach are likely to be small if many projects are developed. Moreover, the bilateral approach allows them to achieve positive externalities such as trade promotion that would not be provided by a multilateral fund. The same applies for governments of big host countries with relevant domestic markets and strong relations to potential big bilateral investors. Big investors from industrial countries are typically emitters, like energy utilities, that face high domestic emission taxes or strong regulation will also lobby for the bilateral approach. They will tend to develop emission reduction projects on their own, because they expect positive externalities to occur and will choose low-risk countries that offer good commercial prospects. They will be interested in creditable emission reduction or sequestration on a short or medium range time-scale. As an international clearinghouse will increase transaction costs, big investors that will prefer a pure project exchange will reject it.

Estimated Demand for JI and CDM Projects

Haites (1998), Vrolijk (1998), Austin *et al.* (1998), Victor *et al.* (1998). and Figueres (1998) have collected estimates of annual demand for reductions during the commitment period for the total Annex B that range between 462 and 1350 Mt C. The supply of "hot air" is seen between 3-1117 Mt – the former being the official Russian estimate, the latter an estimate of International Institute for Applied Systems Analysis (IIASA) (Victor *et al.* 1998). One can find values all over this extremely wide range. Low-cost JI potential in economies in transition is seen at 185 to 305 Mt C while the models see JI and trading around 100 Mt. The range is also wide for CDM market share which is seen between 19 and 57% of the total flexible instruments, ranging from 67 to 723 Mt C. Prices per ton would range from 13 to 42 $ and annual financial flows from 2.8 to 17.4 billion (see Table 2).

Table 2: Estimates for the Potential of the Flexible Mechanisms

Annex B reductions from business-as-usual	580 to 1350 Mt C; median around 1000
"Hot air"	3 to 1117 Mt C
JI and other emissions trading	64 to 110 Mt C, median around 100
CDM	67 to 723 Mt C, median around 400
CDM market share	19 to 57%
CDM prices per ton carbon	13 to 42 $, median around 20
CDM total financial flows	14 to 85 billion $

Sources: Austin *et al.* (1998), Haites (1998), Vrolijk (1998), Figueres (1998), Victor *et al.* (1998).

Who will get these flows? In a CDM without sinks China would be the main beneficiary, receiving 57-70% (mean 63.8%), followed by India (7-14%) (Austin *et al.* 1998).

Rigid Ceilings for Both Instruments

The European Union (EU) and many environmental NGOs have lobbied hard for a strict definition of supplementarity through fixed ceilings on all flexibility instruments. There are different possibilities to implement a ceiling and the proposals discussed so far have been very unclear as to their exact implementation, especially for countries with stabilization or growth targets. The ceiling could be defined as follows:

- x% of the emission reduction target excluding countries with stabilization or growth target (Rule a)
- x% of the respective emission budget (Rule b))
- x% of the average target of Annex B countries (5.2%) (Rule c))
- x% of the reduction from a business as usual path (Rule d))
- Every imported emission permit will be discounted by x% (Rule e))
- A mix of the above such as the EU Proposal of May 1999 that allows the choice between two formulae.

Gylvan Meira Filho (1998) argues for rule b) with x=2.5. Haites (1998) calculates that under x= 49 rule a) would lead to a restriction of the CDM to 30 Mt and of JI to 10 Mt. Rule d) would allow a CDM of 260 Mt and JI of 70 Mt.

Moreover, it is unclear whether the ceiling applies to all flexible instruments together or whether specific ceilings apply for each instrument. Figueres (1998) calculates a CDM-specific ceiling with x=49 under Rule d) that would lead to a CDM of 490 Mt and JI of 250 Mt. Any quotas would lead to an effort to import as much cheap JI and CDM credits instead of selling domestic assigned amount that can be banked.

If a ceiling is introduced, quota allocation to private entities has to be as efficient as possible. Several procedures concerning the activities of private entities could be chosen:

- "First come, first serve". Companies can buy permits until the ceiling is filled. Afterwards, permits can still be bought but only be used in the next commitment period. CDM projects would be advantaged as credits already accrue from 2000. Nevertheless, the following problems could arise: Credits from a JI/CDM project accrue only after the quota is filled. All credits from JI/CDM projects of one investing country could loose their value if the quota had already been filled through emissions trading. This procedure would disadvantage long-term CDM and JI projects and projects with long gestation periods.
- Discounting proportional to the demand surplus. Companies can buy permits until the end of the commitment period. Then the government calculates the aggregate amount of acquired permits – provided they contain the name of the issuing country and project. If it surpasses the ceiling, every permit is discounted according to the demand surplus, e.g. if the quota is 1 Mt of carbon and the permits bought amount to 2 Mt, each permit is discounted by 50%. This method would lead to a high risk concerning the real price of the acquired permits as it is only known after the commitment period and depends on decisions of other companies. Thus risk-averse companies will not invest in permits. Allowing banking part of permits for the next commitment period instead of discounting could reduce the uncertainty. These banked permits would get preference in filling the next quota. Projects with long duration would thus be penalized less.

- Discretionary allocation of the quota according to criteria such as positive externalities, degree of innovation of the projects, diversification of sources of permits etc. Transaction costs, intransparency and uncertainties would be high.

All these allocation modes would disadvantage the project-related mechanisms CDM and JI. They could be attenuated by setting a "soft" quota, which slowly discounts the carbon credits achieved beyond a certain percentage of the domestic target. Any credit beyond another, higher percentage will still be accounted for at a minimum rate. Obviously, this mechanism would not fix the ceiling at an exact value. But domestic reduction would be promoted while the global reduction would be enhanced. But even better would be to discard the idea of ceilings completely.

Efficiency Properties of Different Forms of the Clean Development Mechanism

There are two general options for project-oriented emission reduction: – bilateral and multilateral.[3] The bilateral option allows countries to negotiate a framework agreement setting criteria and rules for crediting. Projects are negotiated freely between entities of both countries.

In the multilateral option investing countries make contributions to an independent fund. Other countries can now offer projects and so compete for the fund's resources. Projects are selected according to their emission reduction efficiency, with positive externalities being taken into account in the case of equally efficient projects. For the duration of the project, each investor country receives a credit proportional to its share of the project portfolio. Project risks would also be pooled with the investor countries being required to pay a corresponding insurance surcharge. The necessary verification could be carried out multilaterally or by private auditors (Mintzer 1994:46 under the term "mutual fund").

The initiators of the CDM proposal clearly envisaged a single multilateral fund. Its efficiency properties will be discussed below compared to bilateral solutions. Intermediate solutions such as a clearinghouse or an information exchange are also covered.

---

3 There also exists a purely unilateral option – i.e. a non-Annex I country finances projects on its own, has emission reductions certified and sells them to Annex I countries. As this needs no specific institutions, I do not discuss it further.

Multilateral Fund

Compared to bilateral CDM, the CDM fund would have efficiency advantages in the following fields: Efficient emission reductions appear possible in principle since projects from any country can be selected. The multilateral approach spreads project risks among all the investors, thus giving even conservative investors and investors with little capital a chance to participate. Transaction costs can be reduced significantly, as the CDM has a much steeper learning curve due to the great number of projects than individual investor countries. The Global Environmental Facility's (GEF) experience in climate protection projects and the World Bank group's accumulated expertise could facilitate the selection and evaluation of projects if the CDM was situated at the World Bank. Then the CDM could also use the preparation done to set up the Prototype Carbon Fund. The concerns of developing countries using these structures could be alleviated through allowing for a double majority voting system right from the beginning.

A multilateral fund could be less efficient than bilateral CDM projects due to the different preferences of individual industrial countries as well as "rent seeking"[4] on the part of the host countries. Given a multilateral solution, it is not possible for the investor countries to select projects according to their own preferences. This, together with the likelihood of large-scale projects being preferred because of their lower administrative costs, reduces project variety. Moreover, the incentive for project partners in the host country to minimize reduction costs would be lost as the difference between the price of emission reductions negotiated by the CDM and the investor country and the corresponding marginal reduction costs accrues to the CDM. A bilateral solution, which generates greater identification with the project in hand, would encourage technological innovation and provide stronger profit incentives for the host party.

Furthermore, there is a danger of institutional inefficiency similar to that witnessed in some subsidiary organizations of the UN. As the CDM has a guarantee that the investors cover its administrative costs, its incentive to keep these costs low is very small. Moreover, the diversion of a share of project revenues for climate change adaptation measures raises the cost of JI and thus lowers global abatement efficiency. To raise these funds, several possibilities exist. The CDM could deduce a fixed percentage of the investors' payments to raise these funds.

Some of the described barriers to an efficient functioning of a CDM fund can be overcome by designing it properly. Salaries of CDM officials could be linked to performance. Transparency should be ensured through third party auditing by accountancy firms, technical surveillance bodies or environmental associations and publication of project data.

The CDM should develop a standard contract which simply require the addition of specific data and regulations for individual projects. This would on the one

---

4 "Rent seeking" is the economic term used for the attempt to retain monetary resources without offering an economic service in return.

hand ensure that the project runs as smoothly as possible, and on the other hand keep the transaction costs down. Free amendment agreements for each of the various sectors are conceivable here for example.

The CDM should also streamline verification procedures by having all projects verified through independent auditors. An expert panel of the UNFCCC Subsidiary Body for Scientific and Technological Advice (SBSTA) could then do spot checks.

Clearinghouse

Besides operating as fund, the CDM could also work as international "clearinghouse" that would accept and evaluate project proposals and invite tenders for projects. This approach differs from the fund approach in that projects are not bundled together in a portfolio. Invitations of tenders are posted world wide and investors can then submit applications. The emission reductions are credited to the successful applicant's home country (Hanisch 1991, Mintzer 1994:46 under the term "managed market"). A large-scale project could possibly be split into several lots.

Compared to bilateral JI, the administration costs generated by a central institution are more than compensated for by the potential investors' individual cost advantages. The costs of locating suitable partners and information costs are much lower than searching on an individual basis and also reduce market entry barriers. The administration costs of a project, which are shared proportionally by the project partners, will also fall as the number of projects and participants increases. Smaller projects, where administration costs form a large proportion of the total costs, would benefit in particular. Project brokerage and placement must not be overly restrictive or complicated. The clearinghouse can help to attract additional project hosts and strengthen basic confidence in CDM projects.

The CDM could set a minimum price per ton of greenhouse gas prevented. The difference between this sum and a project's actual cost would be used to finance administrative costs and adaptation projects. Fixing a price in this way could also be intended to prevent host countries offering projects at dumping prices (Sanhueza *et al.* 1994:17). This assumption disregards economic calculation; host countries will then propose only projects whose declared reduction costs are equal to the minimum price. The difference between the minimum sum and actual costs then accrues to the host country itself. A further characteristic of this concept is that below the minimum sum there is no longer any incentive for investors to carry out CDM projects at all. It is, therefore, a covert quota for emission reductions in the investing country since reduction activities with lower costs per ton than the minimum sum are only carried out at home. Thus, a minimum price should not be set.

Project Exchange

The leanest option for the CDM would be a project exchange where any interested party could gather quick, extensive information on all the CDM projects currently available as well as on corresponding financial opportunities for funding the projects. The projects are all collected in an international database, access to which via Internet is free of charge (Mintzer 1994:46, who gave this model the delightful name "Hackers' Delight"). The participants pay a fee for successful matching to cover costs and raising adaptation project funds.

If the CDM chooses the clearinghouse or project exchange option, it would have to supervise verification rather closely. It should set binding verification standards and accept project proposals only if an independent auditor has already been contracted. The CDM could implement spot checks itself or rely on the SBSTA.

The Baseline Issue

Under both project-related mechanisms emission reductions can only be calculated from a reference basis of emissions, the baseline. An overall definition of a baseline would be the emissions level if the project had not taken place. Art. 6 and 12 of the Kyoto Protocol state that reductions in emissions shall be additional to any that would occur in the absence of the project.

By definition, a baseline cannot be observed and thus cannot be proved to be correct. The aim for climate policymakers should thus be to arrive at a consensus on sensible rules for baseline-setting that build upon criteria derived from a set of policy targets. Then it can be tested whether a proposed baseline has been set up according to the rules.

On the one hand, baselines shall be set in a way to prevent fictitious emissions reductions that would lead to exaggerating the emission budget of Annex I countries. On the other hand, baseline setting shall not reduce efficiency. It will not be possible to reach both targets perfectly – there is an inevitable trade-off!

Investors and hosts of CDM projects – companies as well as countries – want to get a maximum certified emission reduction through the project. They may therefore overstate the possible emission reduction by setting a high baseline. In the case of JI the host country will not have an incentive to do so as the credited emission reduction from JI will be subtracted from its domestic budget. Any overstated baseline will thus lead to the need for higher domestic abatement to be in compliance.

Critical Parameters for Baseline-Setting

There are three critical parameters for baseline setting: determination of additionality, of leakage and lifetime of a project.

The economic additionality of a project – determining whether it has positive incremental costs – is the most difficult issue in the context of baseline determination and has lead to a heated debate (see Baumert 1999 and Rolfe 1998 for an overview). Additionality can be seen on two levels – a macro and a micro level. Due to externalities, they will differ. A project that is clearly additional from a micro-economic point of view may not be macro-economically additional. Under fossil fuel subsidies, for example, a wind power plant might be clearly additional due to higher costs compared with the subsidized fossil fuel. If the subsidy was phased out, it could become non-additional. Thus non-additionality on a macro-level will enhance the supply of micro-level additional projects while strong macro additionality will reduce it.

Micro-economic additionality could in theory be measured according to the following criteria (see Figure 1). They assume that the discount rate and the degree of risk are known, which allows a calculation of risk-neutral costs[5] (for a nice discussion of the effect of different discount rates see Varming *et al.* 1998):

1. Accept all projects that reduce or sequester emissions (as argued for by most of the business community and succinctly stated by Rentz 1998).
2. Prove that the project removes barriers. A list of "accepted" barriers could be defined (IEA 1997).
3. Prove that the internal rate of return (IRR) of the project is lower than that of a commercial alternative.
4. Prove positive "incremental" cost of the emission-reduction related part of the project similarly to procedures used by the Global Environment Facility. The loosening of these procedures in 1998 show that they have been extremely difficult to apply.
5. Prove positive costs of the full project (e.g. by investment modelling) (Bedi 1994, Philibert 1999).

---

5 Obviously, this is difficult to achieve as discount rates will be different from country to country and perceptions of risk are highly subjective.

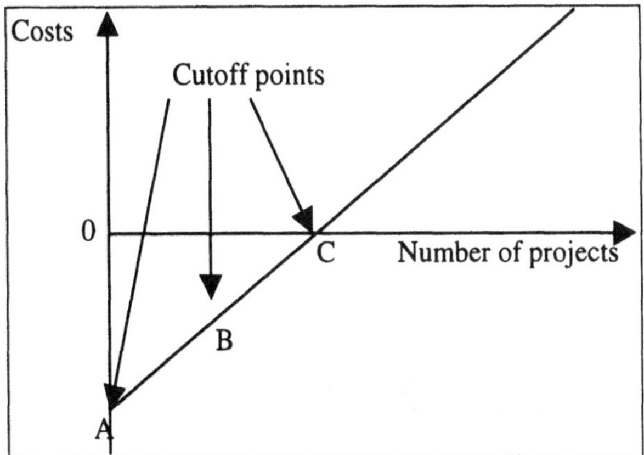

A: All projects that reduce greenhouse gas emissions compared to status quo
B: All projects above a negative cost threshold to account for non-monetary barriers or alternative rates of return
C: Only positive cost projects are accepted

Figure 1: Determination of the Economic Additionality of a Project

Determining micro-economic additionality may be impossible due to a high cheating potential. A narrow approach might lead to the choice of marginal technologies that are not always appropriate and depend on having an undistorted market, which often is not the case in host countries. Experience from AIJ shows that autonomous technology shifts depend on hard-to-observe parameters. On the other hand, macro-economic additionality might be easier to assess. Such an assessment would also have the advantage that there are no perverse incentives to prolongate inefficient policies. The best approach would be to phase in strong macro additionality rules over a certain period of time, e.g. 5 years, to allow countries to change policies.

Leakage determination quantifies positive and negative indirect influences on emissions elsewhere such as technology spillovers (positive) or emissions increases through reduction of market prices of services or commodities that leads to higher demand (negative). The higher the degree of spatial aggregation of a baseline, the more leakage will be covered by the baseline.

Lifetimes of a baseline (see Figure 2) can range from the full project duration ("static" baseline) over revision in case of major "surprises" (i.e. policy or economic shifts) to periodic revision or an ex-ante limited life. Revisions might change the slope or ratchet down a horizontal baseline. The more frequent the revisions and the lower the lifetimes, the higher are costs and risk for project participants. On the other hand, revisions allow to reduce uncertainties that grow with the duration of an unrevised baseline.

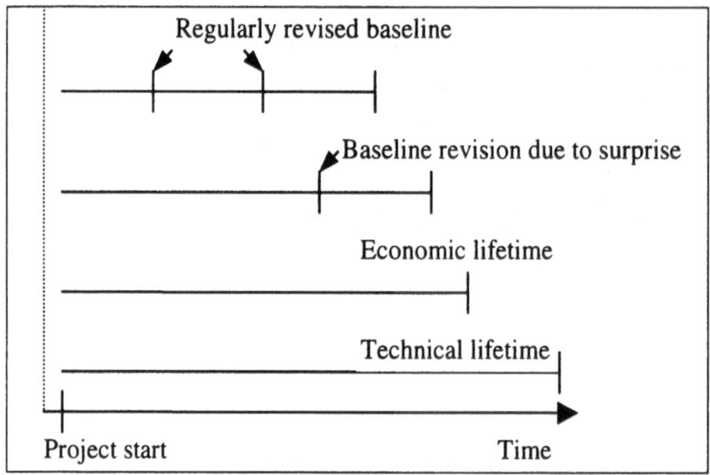

Figure 2: Lifetimes of Baselines

Lessons from Practical Experiences in Baseline Development during the AIJ Pilot Phase

In the past five years, 113 AIJ projects have been approved world wide, and a sizeable share has already been implemented. Therefore, the theoretical debate on baselines can be supplemented by examples from "real life". Nevertheless, part of the database is inadequate, and remains to be improved before the end of the pilot phase. Generally, project-related baselines were used. Often, the baseline methodology is not explained in detail.

Some of the national AIJ programs developed criteria for baseline definition. The criteria of the US Initiative on Joint Implementation (USIJI) are the most detailed ones (Carter 1997). They state that baselines have to be consistent with:

- prevailing standards of environmental protection in the host country
- existing business practices within the particular sector of industry
- trends and changes in these standards and practices

They also stipulate that baselines must include indirect effects such as activity shifting, price effects, and life-cycle effects in products, and that they provide information on other environmental effects of the project.

Nevertheless, USIJI project baseline development was of highly uneven quality and rarely managed to fulfill any of the criteria set above. Indirect effects have not been covered to any extent. For example, renewable project baselines did not include life-cycle emissions of the plant material. Changes in the legal framework were covered only in some projects. Others did not take them into account.

Additionality determination was not really tackled in the AIJ pilot phase with the exception of some Dutch and Norwegian projects (which showed that several projects were clearly non-additional under some definitions discussed above). The baselines of all current projects do not include negative cost-options. Many Swedish small-scale boiler conversion projects in the Baltic states take the status quo before project implementation as baseline. This does not take into account subsidies and market distortions. A phase-out of the subsidies would make many of these projects profitable and thus non-additional if strictly defined. A similar situation applies in the case of the co-generation project in Decin in the Czech Republic which projects a decrease of heat demand by 13% in 2001 and a constant demand thereafter. The existing coal-fired power plant was taken as baseline for heat production. Moreover, the existing average emission factor of the Czech electricity production was taken as baseline for the electricity production of the new plant. That seems to be over optimistic as this emission factor will surely be reduced in the business-as-usual case because of reduction of subsidies.

An even more distorted situation exists in the case of the USIJI RUSAGAS project which entails sealing of valves on natural gas pipelines, takes current emissions as the baseline and estimates a lifetime of 25 years. This baseline clearly shows the importance of micro- versus macroeconomic additionality. So far, the Russian gas company is paid only for the quantity of gas extracted but not for the quantity delivered. If the latter situation applied because of regulatory changes, the incentives to seal the valves would be very high for the company. That means that the baseline would then have to be set to zero.

An interesting consequence of an overly strict definition of macroeconomic additionality occurred in the case of the renewable energy projects in Costa Rica. Due to the Costa Rican Government's commitment to phase out fossil fuel electricity production by 2001, the baseline is zero emissions after 2001. USIJI and independent observers doubt whether the government commitment can be fulfilled, but nevertheless required the baselines to take this commitment into account. This means that renewable energy projects in Costa Rica will not become creditable under the CDM regime after 2000. Therefore, all the renewable energy projects now approved are certainly profitable, thus micro-economically non-additional projects.

To sum up: most of the AIJ pilot phase baselines did not cover indirect effects. Their treatment of the additionality issue was haphazard. Lifetimes were chosen arbitrarily (Ellis 1999). Especially in countries experiencing a rapid change from distorted to deregulated energy markets project baselines have had relevant shortcomings.

Simplified Methods – No Panacea

Due to the experience from the AIJ pilot phase, many researchers (Rentz *et al.* 1998, CCAP 1998a, 1998b) have argued for the development of highly-aggregated baseline methodologies that involve modelling or setting of quantitative, performance-related benchmarks. While reducing the costs for the end user, their development would need a huge public investment and sufficient human capacity. While aggregation might lower the potential for cheating by individual project participants (micro cheating), it could lead to cheating in the choice of parameters for modelling (macro cheating). The aggregated approaches still need a decision about spatial and temporal degree of aggregation (see Figure 3 on the latter).

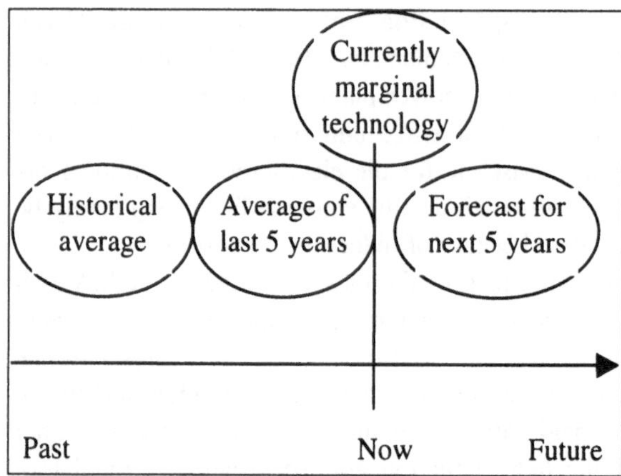

Figure 3: Temporal Dimension of Benchmarks

To capture the differences of specific technologies and the spatial problems described above a matrix of countries and benchmarks/emission factors for a set of baseline technologies could be defined (Jepma 1997). This could be even more simplified by using project categories instead of technologies (Michaelowa 1998, Michaelowa & Dutschke 1999). The lowest degree of aggregation would be standardized parameter setting for project baselines (Begg *et al.* 1999). This would include rules to ensure environmental integrity such as monitoring, baseline revision, limited crediting life. Matsuo (1999) suggests a stepwise standardization which starts from purely project specific baselines and tries to use the experience to define project categories that can use the same methodology.

It seems that depending on the situation and project type there are different types of "optimal" baselines:

- Forestry, infrastructure, policies, large number of projects in all sectors: highly aggregated, benchmarks
- Large projects, many projects in a specific sector, fuel substitution: sector-specific, technology or default matrix, project-related standardization
- Small projects such as renewables, retrofits, small number of projects: project-related standardization, technology or default matrix

To test what approach is most appropriate it might be useful to let baselines compete. This might lower transparency, though and risk "crowding out" of good baseline methodologies by bad ones. A possible solution might be: As long as COP has not decided on final rules for CDM, one could test different approaches in practical project settings and simulation on the basis of existing (AIJ) projects. It would have to be made sure that no methodology would be unfairly penalized after the decision has been taken.

The final decision might decide on different baseline methodologies for different situations. The decision could use thresholds to determine the change from one to another baseline methodology such as:

- If the share of CDM project investment in the annual investment of a country exceeds x%, the baseline methodology shall be switched to a country-/benchmark approach, or
- If the share of CDM project investment in the annual investment of a country's sector y exceeds x%, the baseline methodology shall be switched to a sector-specific approach

An alternative might be that the decision allows the CDM Executive Board to set such thresholds and to develop the methodology in detail.

Options for Future Development

The differences between CDM and JI could be reduced by decisions taken at COP 6. In fact, there are some tendencies to align the two mechanisms:

Extension of the Adaptation Tax to All Mechanisms

Many NGOs have argued for an extension of the adaptation tax to all flexible mechanisms. These demands have been taken up by representatives of developing countries. Recently, they have got support from the research community. Grubb *et al.* (1999:222f.) argue for a tax of 5 $/t on all transfers of emission permits. Any such proposal, however, will encounter strong opposition from both prospective buyers and permit sellers under Art. 17.

Early JI to Reduce "Hot Air"

Principally, early JI would work as follows (see Figure 4): Early JI credits are given to investors in form of futures on the first budget period where they have to be subtracted from the host country budget. Otherwise the host country would have an incentive to maximize early JI that is not additional and thus the Annex B budget would be blown up. If JI leads to a reduction from business-as-usual by the amount A, the same amount has to be deducted from the host country budget to avoid non-compliance.

In this case, one might think that JI reduces the amount of "hot air" by the credited emissions reduction. This was done by the Swiss delegation at the fourth Conference of the Parties in Buenos Aires which circulated a non-paper that argued for early crediting of JI exactly for this reason (Switzerland 1998). This argument holds if "hot air" is defined as the amount of allocated permits exceeding the initially forecast emissions for the budget period. However, if only domestic efforts are considered to determine the amount of "hot air" – what in my view is more appropriate – , the business-as-usual emissions path has to be adjusted due to JI activities since JI should be classified as a non-domestic effort – JI is carried out in the host country but financed by the investor country.

With this definition in mind, a reduction in "hot air" through early JI would require A>B in Figure 4. This depends on baseline setting, induced change in the future emissions path and especially on the start of the program. Therefore, we would argue that the amount of "hot air" is likely to increase (A<B) if early JI is truly additional as the business-as-usual path is shifted downwards. Furthermore, if the program starts before the year 2003 it runs longer than the first commitment period resulting in a longer period of time where early credits can be accumulated.

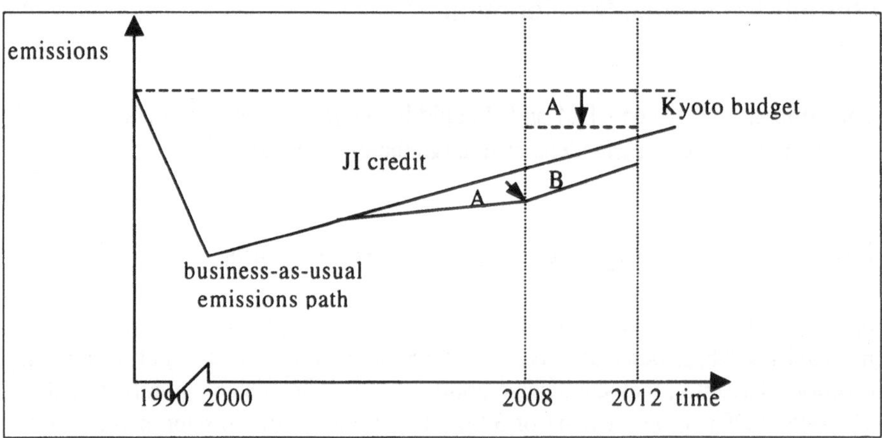

Figure 4: JI Early Crediting in a Host Country with "Hot air"

Only in case of non-additional JI the amount of "hot air" – taking our definition into consideration – seems to fall but this is only due to a "laundering" of the "hot air". Non-additional JI would not be attractive as it would reduce the saleable quantity of "hot air" by the same amount and presumably entail lower transfers.

Notwithstanding the design of such a program or whether the amount of "hot air" will decline or not – as long as there is a subtraction from the host's target, the total Annex B budget remains unchanged.

Recognizing the problems of the earlier approach, Switzerland further developed its proposal for the June 1999 session of the climate negotiations by stating that revenue from early JI credit sale should be reinvested in emission reduction projects (UNFCCC 1999).

Conclusions and Recommendations for Future Negotiations

The flexible instruments of climate policy have to be both statically and dynamically cost-efficient. This is not assured as the Kyoto Protocol stands today. While it allows project-related crediting of emissions reduction abroad, it has set up different mechanisms – JI for projects among industrialized countries and the CDM for projects with the rest of the world. There are substantial differences in the rules for these instruments. A major one is that CDM credits accrue from 2000 while JI is only credited from 2008. The CDM is subject to an adaptation and administration tax while JI is not. CDM credits come into being only after certification through an independent body. While the CDM will be supervised by some kind of multilateral organization, no such organization is foreseen for JI. Two rules potentially explosive for JI state that credits only accrue if reporting requirements are met and there are no official doubts about compliance. The exact impact of these differences cannot be estimated today as many of the rules remain to be clarified. It is clear, though that there is no obvious bias in favor of one instrument. Applying the same rules to both mechanisms would eliminate any bias. Indeed there are tendencies in the climate negotiations to move into this direction.

It would be advisable to use both a multilateral and bilateral CDM simultaneously as each has advantages for certain constituencies. The competition of both modes would reduce transaction cost. Concerning crediting, it would be advisable not to set ceilings for CDM and JI investment. Development of and decisions on baseline methodologies will be a crucial point that might determine the competitive position of CDM and JI towards emission trading.

References

Austin, D., Seroa da Motta, R., Zou J., Li J., Pathak, M. & Srivastava, L. (1998), *Opportunities for financing sustainable development via the CDM.* Buenos Aires.

Baumert, K. (1999), *Understanding additionality.* In Goldemberg, J. & Reid, W. (eds.), Promoting development while limiting greenhouse gas emissions: trends & baselines. New York, 135-143.

Bedi, C. (1994), *No-regrets under Joint Implementation?* In Ghosh, P. & Puri, J. (eds.), Joint Implementation of climate change commitments. New Delhi, 103-107.

Begg, K., Parkinson, S., Jackson, T., Morthorst, P.-E., & Bailey, P. (1999), *Overall issues for accounting for the emissions reductions of JI projects.* In NEDO/GISPRI (eds.), CDM workshop – workshop on baseline for CDM, Proceedings. Tokyo, 167-190.

Carter, L. (1997), *Additionality: the USIJI experience.* Paper presented at the Workshop on environmental benefits of AIJ, 9-10 September, Paris.

Center for Clean Air Policy (1998a), *Emission caps and methods to quantify project emission baselines.* Washington.

Center for Clean Air Policy (1998b), *Top-down baselines to simplify setting of project emission baselines for JI and the CDM.* Washington.

Center for Clean Air Policy (1998c), *JI for credit now: Establishing early Joint Implementation programs.* Washington.

Ellerman, D. & Decaux, A. (1998), *Analysis of post-Kyoto emissions trading using marginal abatement cost curves.* Boston.

Ellis, J. (1999), *Experience with emission baselines under the AIJ pilot phase.* OECD Information Paper, Paris.

Figueres, C. (1998), *How many tons? Potential flows through the Clean Development Mechanism.* In WRI, FIELD, CSDA (eds), The Clean Development Mechanism. Washington, 19-22.

Greenpeace (1998), *Making the Clean Development Mechanism clean and green.* Buenos Aires.

Grubb, M., Vrolijk, C, & Brack, D. (1999), *The Kyoto Protocol.* Earthscan, London.

Gylvan Meira Filho, L. (1998), *Ideas for implementation.* In Goldemberg, J. (ed.), The Clean Development Mechanism: Issues and options. New York, 39-43.

Haites, E. (1998), *Estimate of the potential market for cooperative mechanisms 2010.* Mimeo, Toronto.

International Energy Agency (1997), *Activities Implemented Jointly — Partnerships for Climate and Development.* Paris.

Jepma, C. (1997), On the baseline. *Joint Implementation Quarterly* **3**(2), 1.

Matsuo, N. (1999), *Step-by step standardization for baseline setting.* Paper presented at SBSTA 10. Bonn.

Michaelowa, A. (1998), Joint Implementation – the baseline issue. *Global Environmental Change* **8**(1), 81-92.

Michaelowa, A. & Dutschke, M. (1999), *Economic and political aspects of baselines in the CDM context.* In Goldemberg, J. & Reid, W. (eds.), Promoting development while limiting greenhouse gas emissions: trends & baselines. New York, 115-134.

Michaelowa, A. & Schmidt, H. (1997), A dynamic crediting regime for Joint Implementation to foster innovation in the long term. *Mitigation and Adaptation Strategies for Global Change*, **2**(1), 45-56.

Mintzer, I. (1994), *Institutional options and operational challenges in the management of a Joint Implementation regime.* In Ramakrishna, K. (ed.), Criteria for Joint Implementation under the Framework Convention on Climate Change. Woods Hole Research Center, 41-50.

Philibert, C. (1999), *Think piece on the additionality under the CDM.* Paris.

Rentz, H. (1998), Joint Implementation and the question of additionality — a proposal for a pragmatic approach to identify possible Joint Implementation projects. *Energy Policy* **4**, 275-279.

Rentz, O., Wietschel, M., Ardone, A., Fichtner, W. & Göbelt, M. (1998), *Zur Effizienz einer länderübergreifenden Zusammenarbeit bei der Klimavorsorge.* Karlsruhe.

Rolfe, C. (1998), *Additionality: What is it? Does it matter?* Vancouver.

Sanhueza, E., Van Hauwermeiren, S. & De Wel, B. (1994), *Joint Implementation: conditions for a fair mechanism.* Instituto de ecologia politica, Santiago de Chile.

Switzerland (1998), *Initial ideas on pre-2008 joint implementation.* Mimeo, Buenos Aires.

U.N. Framework Convention on Climate Change (1999), *Prinicipies, modalities, rules and guidelines for the mechanisms under Articles 6, 12 and 17 of the Kyoto Protocol.* Submissions from Parties, Paper No. 3 Switzerland, FCCC/SB/1999/MISC.3/Add.2/Corr.1, Bonn.

U.N. Framework Convention on Climate Change (1997), *Kyoto Protocol to the United Nations Framework Convention on Climate Change.* FCCC/CP/L.7/Add.1, Kyoto.

Varming, S., Larsen, P. & Christensen, B. (1998), *Possibilities of AIJ: two Polish cases.* In Riemer, P., Smith, A. & Thambimuthu, K. (eds.), Greenhouse gas mitigation. Technologies for Activities Implemented Jointly. Amsterdam, 529-534.

Victor, D., Nakicenovic, N. & Victor, N. (1998), *The Kyoto Protocol carbon bubble: implications for Russia, Ukraine and emission trading.* IR 98-094, Laxenburg.

Vrolijk, C. (1998), *The potential size of the CDM.* Mimeo, London.

Key Developments Related to Greenhouse Gas Trading and Key Issues to Be Resolved

Viktor Danilov-Danilian

Russian Federation State Committee on Environmental Protection,
c/o Vitaly Gorokhov, Höhefeldstr. 31, D-76356 Weingarten, Germany,
e-mail: vitaly.gorokhov@t-online.de.

First of all, what is the principal reason for our concern? No doubt, we are concerned about negative changes of global climate. We are able to predict now only some of the possible consequences of global climate change, their negative impact on the national economies and world's economy. We also realize that numerous island countries may be completely wiped out of the face of the Earth.

Do we have credible scientific data to substantiate these apocalyptic predictions? Yes, we do. We carefully analyzed chemical content of the Earth's atmosphere for the last 200 million years and correspondingly the dynamics and intensity of change of chemical content of the atmosphere during this period. We are quite positive now about the unprecedented rate of increase of Carbon dioxide (CO_2) concentration which is recorded during the last 100 years. Such a change was never recorded in the past 200 million years. The reason for such a change is obvious: It is anthropogenic emission of carbon dioxide, which is accompanied by emission of other by-products of technogenic civilization, including several greenhouse gases (GHG).

Are these changes dangerous indeed? There are many answers to this. The most common answer stems from the well-known "greenhouse" model of the Earth's climate. This model, I think, needs to be carefully examined, but it is not quite convincing. To say the truth, no one can prove yet that the Earth's climate changes according to the greenhouse model. We need at least another ten years of observation and modeling to be quite positive about the nature of global climate change.

But I am hundred-per-cent sure that during the last 100 years the Earth's atmosphere and climate deviated from their equilibrium states. The supporting back-of-the-envelope calculations were presented in numerous scientific reports. In the past, both biota of oceans and surface biota were net absorbers of CO_2. Now only the oceans retained this function. Surface eco-systems (including humans) now emit more CO_2 than they absorb. Worse of all, anthropogenically changed eco-systems (or exited eco-systems, as physicists say) have become net emitters of CO_2.

These deviations from equilibrium state are quite hazardous. Irrespective of the particular scenario of global climate change, human civilization will have to cope with catastrophic consequences of equilibrium loss, if this process will not be stopped. The loss of global equilibrium is a matter of grave concern, even if greenhouse model should be replaced by some other yet unknown scientific paradigm. It is imperative that the causes of global climatic equilibrium loss should be blocked away.

The principal aim of Global Climate Change Convention is restoration of balance of chemical composition of the Earth's atmosphere. Around this ambitious task were centered numerous international debates held in preparation of the Convention for the Conference in Rio and subsequent events – Conferences in Berlin and Kyoto, Buenos-Aires and Bonn.

Obviously, climate change is a global problem in all its aspects. First, no one will avoid the consequences of Earth's climate equilibrium loss. Second, all countries, all people triggered this profound change. We, humans, are causes and victims of this complicated process. Third, this problem can be solved only internationally. It requires global effort and global co-ordination.

One has to admit regretfully that humanity is yet to learn how to initiate global effort and co-ordinate it. With numerous global problems we face now, there is no uniform mechanism have been developed to solve them. Our efforts and approaches vary greatly with locality and particular context, no matter how one would measure the effort – in costs or results, no matter how one would aggregate it – per square kilometer, per dollar invested, or per capita.

How we should proceed with co-ordination of our efforts to solve international problems? The logic is simple to grasp and difficult to implement. The idea is to assess the resources needed in principle to solve the problem, and then to apportion the resources among the countries according to their economic capacity. Next step is to channel available resources where they can produce most benefit, i.e. to use them most effectively.

According to this elementary logic, one may solve the problem of GHG emission reduction in two steps: the first step is "budgeting", one needs to assess how much resources is required, the second step is "bargaining", where to get the required resources? The answer to the first question is still not quite clear. The second question leaves no doubt: rich countries are to pay. After all, they are relatively more responsible for distorting the Earth's climatic balance. The collected revenues should be directed where one may get the greatest environmental benefit per each dollar spent. Environmental benefit is not difficult to measure: it is GHG emission reduction or sequestration volume. The results of reduction and ecosystem sequestration are equivalent however different the mechanisms are. To achieve emission reduction, certain measures are to be taken in the real sector of economy. To sequester greenhouse gases one needs to increase assimilative capacity of ecosystems, which means restoration of their virgin (natural) state.

It is my opinion that that idea of emission trading remains now the only adequate mechanism of addressing global problems. This idea belongs not only to the environmental field. It may be broadened out to address other global problems as well.

Emission trading is a bright idea which offers a mechanism of practical realization of international co-operation in solving global problems. It has several important advantages. To put this mechanism into work, all countries have to accept certain obligations to reduce their emissions of greenhouse gases, most importantly CO_2. Without binding obligations there will be no emission trading.

What does this mean – to take obligations? This means accepting certain responsibilities with respect to international community. This demonstrates a country's good will. If humans, countries, international community are unable to take this step, our future will be undoubtfully miserable. Our chance to survive on the planet we're only beginning to destroy depends upon the free will to accept such obligations, to take the first step.

This first step immediately translates into purely economic measures. Certain economic targets ensue, certain resources are needed.

Because of global nature of climate change, we don't have to limit countries' obligations to their own territory. Emission reductions and/or GHG sequestration may be achieved in any country, any climate zone. The only requirement should be environmental effectiveness of such measures. The important conclusion is that the country which accepted obligations doesn't necessarily implement them on its own territory.

If we agreed on this, formation of global emission reduction market comes in quite naturally. A country which accepted obligation to achieve emission reductions will have to do so in a least expensive way. This means to identify a measure with maximal environmental effect per dollar invested. This is what market is all about. It solves profit maximization problem every moment it exists.

No doubt, emission reduction quota market has its specific features. First of all, it is international market. Secondly, its effectiveness is measured by environmental indicators. Thirdly, participating countries must enter into legally-binding emission reduction agreements. But the underlying mechanism is market-based. It seems suitable even for *a priori* non-market problems.

This mechanism was formulated in the process of designing the solutions for global environmental problems; it is called "joint implementation" (JI). Emission trading presents a brilliant example of a market-based JI mechanism. This is why I think that emission trading beats all other tangible mechanisms of solving global problems.

The idea of joint implementation is not altogether new. If there's anything new in it – it is the particular field of its implementation: international obligations. If two countries (or more) agreed to accept certain mutual obligations then the easiest way to implement them is to act co-operatively. The individual obligations (e.g. emission reduction targets) should be summed up to the "common benchmark". This idea, however not new, has been rarely realized in practice. World community still has an unused resource of international relations.

Let us take climate change policy as an example. Economists will find this primer quite elementary. I beg their pardon in advance, because my intention is to explain the matter to non-economists who are interested in quota market. Assume country A has to reduce its emissions by a tons, while country B has a target of b tons. Because country A has already invested a lot in energy-saving technologies, any new investments will have very little effect (all low-cost measures have already been taken). Its marginal cost of emission reduction is m dollars per ton. Country B has not invested in energy-saving yet, thus its marginal cost of emission reduction is n, where $n \ll m$.

If the two countries act independently, they will both spend $am+bn$ dollars on emission reduction (for simplicity we assume linearity in calculation of total costs of emission reduction). If the two countries act jointly, all emission reductions will be achieved in country B (where it is cheaper to do), and the total costs will be $(a+b)n$ dollars. The total "savings" with respect to the first scenario will be $a(m-n)$ dollars.

It is likely that country A will choose to spend an dollars abroad instead of spending am dollars home to achieve the same environmental effect. It means that country B will get an dollars in investment, while country A will save $a(m-n)$ dollars achieving its international obligation.

Country B will benefit from the investment because introduction of new technologies will have positive systemic effect on its economy. This is why country B may be willing to participate in the technology transfer. In this case the cost breaks down: country A pays only a fraction k, where $0<k<1$ while country B pays remaining $(1-k)$. This will result in the following break-down of implementation costs: to achieve reduction of emissions by a tons country A will pay kan dollars and country B will pay $(1-k)an$ dollars.

Of course, this would give rise to criticisms in country B: critics will say the developed country A robbed poor developing B which "despite its hardships had to pay $(1-k)an$ dollars to help a rich country A to fulfill its obligations". We leave these claims unanswered. Besides, foreign investment may combine with domestic emission reduction measures in country B (it would have to spend bn dollars to comply with international treaty). In this case country B will spend less than bn, and will attract kan dollars of foreign investment.

This trivial example illustrates one simple but tremendously important point: environmental co-operation (and not only environmental) may be profitable for both countries. However, even with demonstrated good will two problems will remain.

First problem is to create legal basis for JI projects, to "write them up" in the Protocol. This means to allow the parties to the Protocol to achieve their obligations by direct or indirect participation in emission reduction projects abroad.

Global climate will benefit from all such projects no matter where they take place. This is, so to speak, our common endowment. This problem already got attention of legal experts, they provided the solution.

Second problem is to create the procedures for verification of compliance, and for determination of financial burdens for all participating countries. These procedures have to be objective, regular, convincing, and fair. Only market procedures are known to satisfy these requirements. Therefore, we need to put market procedures to work in the above said context. GHG emission trading provides the solution for this problem.

A simplest illustration of emission trading stems from the above example. It may be interpreted as follows: country A instead of domestic implementation finances emission reduction measures in country B. In doing so, country A purchases an emission reduction quota of *a* tons from country B at the price of *kn* dollars per ton. (We ignore a whole range of financing details here).

Other schemes of emission trading may be contemplated as well. For instance, country may have an "emission pool" if its emission cap has not been reached yet. This situation is observed in Russia, Ukraine and some other post-communist countries. Then the difference between the actual emissions and emission limit is defined as quota which may be freely sold out this year. The country-seller however will not be able to use the revenues at its discretion. The revenues from such transactions may be spent only on energy-saving technologies or other GHG-related environmental projects.

Principal actors on the GHG quota market will be enterprises and firms (countries and groups of countries may also participate in principle). It is not clear yet how to translate national emission quotas into obligations of individual buyers and sellers on this market. The restrictions on revenue use should also be specified more carefully. Special monitoring measures are needed to control the actual emission levels and ensure compliance during transactions. Without such monitoring GHG emission trading market will not achieve its principal objective. But most importantly, when procedures and rules are specified, the price of GHG quotas will be determined by the market itself, matching supply and demand.

Some critics of market approach to GHG emission reduction offer imposition of greenhouse emission tax as an alternative approach to climate change problem. This instrument has been thoroughly studied by ecologists and economists. It was discussed during the conference "Environment for Europe" held in 1994 in Lucerne. At first this idea seemed attractive and easy to implement. But carbon tax has not been implemented ever since, and the interest has lowered considerably. What is the reason?

Let us consider carbon dioxide – the most important of greenhouse gases. (Water vapour is not considered here). Other greenhouse gases even taken together pose less global warming danger. Besides, chlorofluorocarbons (CFCs) are regulated by other treaties, nitrous oxides are being taxed separately as toxic substances. Industrial methane is a valuable fuel source and for this reason it is economically viable for industry not to waste it. Agricultural methane on the other hand is difficult to monitor or regulate, this is why it cannot be seriously analyzed in economic context of various emission reduction instruments.

Carbon dioxide emissions are proportional to the amount of fuel burned. This is a remarkable fact: to determine the emission levels one just needs to account for fuel use which is less difficult to do than monitoring emissions. On the other hand, CO_2 emission charge becomes an additional tax on fuels (with all the objections which a newly imposed tax can provoke). Fuel taxation may in principle stimulate energy saving, but this effect is quite small, as the economic analysis of pollution charges showed. Setting emission limits proved to be more effective tool especially when combined with economic cushioning measures. By the way, emission quota market is an example of such a cushion if used in combination with emission caps. These caps were actually set for the parties by accepting legally binding obligations to reduce carbon dioxide emissions.

There is no clear vision of how to use proceeds from carbon tax. Tax revenues should be used for energy saving, but how exactly it should be done? Centralized planning is certainly not an effective mechanism of income redistribution.

Let us compare carbon tax (or environmental tax on fuel use) with emission trading. It is my opinion that in every respect emission trading is preferable.

Indeed, emission trading helps all countries achieve their emission reduction targets. This principal objective is not directly connected to imposition of a carbon tax. The link between carbon tax and specified emission reduction volume is rather tenuous.

Another advantage of emission trading is that the countries-parties to the protocol are free to trade. The principle of voluntariness is very important for international treaties although one cannot completely exclude national ambitions, competition among countries, attempts to exert political force.

Yet another aspect is ability to use market mechanisms. While emission trade scheme offers all advantages of free market operation, carbon tax as a fiscal instrument distorts operation of markets.

I already mentioned the direct link between emission limits and quota trading. Carbon tax by itself will not ensure compliance with agreed emission limits. This compliance may be achieved by a complex feedback mechanism. To set up such a mechanism on international scale, a concerted international bureaucratic effort is needed, involving scores of responsible, honest and qualified public servants. It is extremely difficult to make them accept common rules, procedures, philosophy towards fiscal policies, climate change policies, international disputes.

It is worth adding that a whole army of fiscal officers is needed to collect a new massive tax on a national level. A strong feature of emission trading is its decentralized character. It would take considerably less administration costs. Only a handful of controllers will be needed to monitor the performance of quota market on international level because emission trading will be effectively administered nationally by market forces themselves.

I can hardly think of a single advantage of a carbon tax compared to emission trading. International trade in emission allowances offers a fruitful ground for practical realization of joint implementation mechanism which has been actively propagated on recent international fora. JI mechanism, to my opinion, is an important achievement of our collective thought, it turns a new page in international relations.

Let me point out once again that emission trading does not serve self-centered national ambitions of any single country. It offers a solution for our common problem. This solution incorporates individual interests of all participating countries. Without taking into account economic interests of all countries no international problem can be solved. This is the reality.

There are many unresolved issues concerning operation of emission quota market. Who should be its players? I think that both firms and counties, and groups of countries should participate as buyers and sellers. The less restrictions we impose on market structure the more effectively it will operate. We should strive for maximum economic freedom within the scope of this market.

I, as an economist, have been interested in this problem for a long time, at least for the last thirty years. There was much discussion in the former Soviet Union how to harness market forces to make them work in the command economy. I always pointed out that instead of "harnessing" we should let market forces act freely within well defined economic space.

We should carefully define the scope of the market, and not interfere with its operation within the defined limits. In this presentation, I talked about some of these limits, but not all. Let me stress once again that once the market is defined and set in place, our task will be to let it go, not to help the market forces manifest themselves. I foresee a really free market within clearly defined boundaries. Because market mechanism proved to be the most effective economic mechanism which can make the most use of limited resources, and can achieve tremendous results which no one could have ever expected in advance.

US Experience in SO_2 and NO_x Emissions Trading and Developments in the Greenhouse Gas Market

Garth Edward

Natsource Emissions Brokerage Desk[1], 140 Broadway, 30th Floor, New York, NY 10005, USA, e-mail: gedward@natsource.com.

SO_2 Emissions Allowance Trading Program

SO_2 emissions allowances grant the right to emit one ton of sulfur dioxide (SO_2) into the atmosphere. Allowances are:

- standardized, issued in vintage years
    - tradable, any individual can open a US Environmental Protection Agency (EPA) account
- bankable, unused allowances carry forward
    - issued in three phases
    - issued at zero cost basis to affected sources

---

1 Natsource is a major institutional energy broker headquartered in North America and active in the natural gas, electricity, coal, emissions, and weather derivatives markets. They are the highest volume US broker of emissions allowances, with more than $1.5bn in transactions since the establishment of the markets over the last few years. The client base includes over 400 institutional firms such as utilities, investment and commercial banks, energy producers, insurance and reinsurance companies, energy marketers, agricultural firms, and heavy industry. Natsource is a founding member of the Emissions Marketing Association which now has over 150 members from 80 companies. Furthermore, Natsource is a member of the US Center for Clean Air Policy Greenhouse Gas Braintrust and also serves as an advisor to the Whitehouse Climate Change Taskforce.

The SO_2 market was effectively established by the Clean Air Act, signed 11/14/90, and allowance allocations were finalized on 3/23/93. Allowances are issued by vintage and allowances not used in specific vintage years may be used in future years. First trades took place in 1993 at $250-$300/ton.

The US EPA Acid Rain Division opened its allowance tracking system to register transactions in March 1994. All *transfers* are recorded and posted on the internet: www.epa.gov/acidrain/atsdata. The tracking system does not track forward transactions or options unless executed and transferred.

In addition to standard "over-the-counter" trading, the US EPA holds an annual SO_2 auction. This auction is conducted for the EPA by the Chicago Board of Trade. The purpose of the auction was to ensure new units have access to a public source of allowances and provide price signals to the market. However, these reasons may be less important as the market matures. The auctioned allowances are withheld by EPA in a special auction reserve which constitutes 2.8% of initial annual allocations. The auction has typically constituted much less than 1% of total annual trading activity.

The SO_2 market will expand its scope and increase stringency:

- Phase I ('95-'99) 20% reduction in SO_2 from ('87 levels), 263 units/110 plants
- Phase II ('00→) 50% reduction in SO_2 from ('87 levels), 2,000+ units, all new plants 25MW+

SO_2 market participants include electric utilities, energy marketers/traders coal suppliers, independent power producers (IPPs), merchant plant developers, introductory brokers and small diesel refiners.

Prices in the SO_2 allowance market are driven by:

- cost based drivers
  - coal: cost adders required to bring coals into Phase II SO_2 limits range from -$2/ton to $14/ton
  - oil/gas: fuel switching to lower sulfur fuel
  - control equipment: flue-gas desulfurization (FGD) systems $150/ton++
- regulatory/industry restructuring drivers
  - new regulations for ozone, regional haze, particulate matter ($PM_{2.5}$) and carbon dioxide (CO_2) will impact SO_2 value
  - deregulation forcing mergers/acquisitions, divestiture, utilization and ratemaking uncertainty

US EPA records indicate that a total of 25M tons have been transferred since 4/94. In 1998, there were approximately 1M spot trades and 1.3M option trades. In the first quarter of 1999 there have been 388,000 spot trades and 966,750 option trades.

NO_x Emissions Trading

There are three types of nitrogen oxides (NO_x) markets in the US

1. new source/offset markets
   - permanent emission reduction credits (ERCs)
2. compliance markets – voluntary open markets
   - discrete emission reductions (DERs)
   - verified emission reductions (VERs)
   - "mass" emission reduction credits (ERCs)
3. compliance markets – closed/mandated cap & trade
   - Ozone Transport Region (Commission) – allowances
   - SIP Call Region – allowances
   - Regional Clean Air Incentive Market (RECLAIM) trading credits – RTCs

The most active NO_x market is the mandated cap & trade allowance market established under the Ozone Transport Commission. A 1994 memorandum of understanding sets the NO_x budget (tons) for the Ozone Transport Region (OTR).

NO_x emissions allowances are similar to the SO_2 allowance market. A NO_x allowance provides the right to emit one ton of oxides of nitrogen into the atmosphere from May 1 through September 30. NO_x allowances are:

- standardized, issued in vintage years
- tradable, any individual can open EPA account
- bankable, unused allowances carry forward with restrictions under PFC
- issued in two phases
- issued at zero cost basis to affected sources

NO_x allowances are allocated in two phases (219K '99 - '02 and 143K '03 -) for the ozone season which is specified as May - September. 465 sources are initially affected including utilities, IPPs, and industrials (>250 mmbtu/hr or >=15 MW).

NO_x market participants are similar to SO_2 market participants with the addition of waste-to-energy facilities, chemical plants, refineries, other industry (paper, steel, etc.), and control equipment vendors.

Prices in the OTC NO_x market are driven by:

- cost based drivers:
  - fuel switching and co-firing: low vol coal, gas
  - combustion tuning
  - post-combustion controls – selective non-catalytic reduction (SNCR), selective catalytic reduction (SCR)
- regulatory/industry restructuring drivers:
  - deregulation in Northeast a big OTC driver
  - finalization of some state regulations, allocations

- coordination of OTC and SIP call markets in '03

Original estimates of NO_x allowance prices at $1,500/ton have been exceeded five-fold with a market high at $7,600. More than 35,000 tons have traded to date, with 50-100 ton lots the standard.

NO_x trading has had a significant impact on power prices. Assuming $5,000 allowance, a 10,000 Btu/kWh heat rate, the following are NO_x cost adders for different rates:

- 0.45 lbs./mmBtu results in $11.25/MWh
- 0.30 lbs./mmBtu results in $7.50/MWh
- 0.20 lbs./mmBtu results in $5.00/MWh

Greenhouse Gas Market

There is increasing market activity with increasing numbers of greenhouse gas (GHG) transactions and the development of major portfolios. Early action crediting legislation is under development in US, Canada, Norway, Denmark, and New Zealand etc..

Business and industry does not consider the science of global warming to be certain, nor is there certainty on legislation. However, there is certainty that the threat of emission reduction legislation presents a major business risk. The greenhouse gas market is primarily a risk management market rather than a compliance market at this stage.

The tradable units are metric ton units of CO_2 emissions, or CO_2 equivalent emissions. Emission reduction credits typically meet the following minimum criteria:

- specific and identifiable reduction of emissions
- measurable and verifiable
- ownership clearly established
- surplus, not otherwise required by law
- reviewed/certified by independent third party

At the present stage of legislative uncertainty there are two major risks in GHG transactions, and two major responses have emerged:

- regulatory risk: that no emission reduction legislation enters into force. Market participants combat this risk by using option structures, especially with later expirations.
- validity risk: that credits prove to be invalid in a future GHG emissions trading system, either at the domestic or international level. Market participants combat this risk by negotiating whether buyer or seller should be held responsible for the validity of the credits under a future trading system.

There is now increasing commercialization in the GHG market with participants moving from public relations to comprehensive commercial risk management strategies. There is also a general shift from industry opposition to participation.

The broker has an important role in developing GHG transactions. The broker aims to assist market participants to capture opportunity and manage risk. The broker's role is to:

- provide price discovery
- run a continuous auction, matching buyer and seller
- negotiate commercial terms
- develop transaction structures
- help establish market convention
- provide education and seminars

Political and Economic Scope for Permit Markets in Europe

Gert Tinggaard Svendsen

Department of Economics, The Aarhus School of Business, Fuglesangs Allé 20, DK-8210 Aarhus V, Denmark, e-mail: gts@hha.dk.

Introduction

What kind of design allows society to achieve environmental objectives cost-effectively? In general, economists have not considered political concerns and second-best world solutions in the traditional analysis of economic instruments. The contribution here is to find a cost-effective approach to environmental regulation in Europe and thereby diminish the gap between theory and practice.[1] If a policy is not designed in a politically acceptable way, it will inevitably be changed away from its cost-effective design during the political decision-making process because none of the dominating interest groups like it. There must be something in it for them.

Up until now, the use of taxation (and standards) dominated the environmental policy scene in Europe (Howe 1994). It seems that there has been (and still is) great reluctance to use tradable permits in the regulation of environmental problems – especially among the European countries. However, the newest US experience with permit markets has been very positive. This is due to 'grandfathering' meaning that regulated parties are given emission rights for free, typically according to their historical emissions. In the following, the term 'permit market'

---

1 This gap has been highlighted by Green & Shapiro (1994), Hahn & Stavins (1992) and Svendsen (1998a and b). See also Tietenberg (1985), Cropper & Oates (1992) and Baumol & Oates (1988) for nice overviews concerning the use of economic instruments.

refers to this US experience of 'grandfathered' permit markets where property rights to historical emission rights are transferred for free.[2]

Note that Carbon dioxide (CO_2) emissions cannot increase if the CO_2 quotas are distributed based on historical emissions. Each participating country is given a CO_2 quota corresponding to its 1990 emission. The number of CO_2 quotas in circulation is "frozen" in this way. The principle corresponds to distributing fishing quotas to secure a constant fish stock. Next step is to devalue the quotas of the individual countries.

For example, the United Nations (UN) climate agreement in Kyoto says that the European Union (EU) must reduce its emissions of greenhouse gases by 8%, the US by 7% and Japan by 6% from 1990 to 2012. Industrialized countries as a whole must reduce their emissions of greenhouse gases by 5.2%. So devaluation of quotas, following the Kyoto agreement, can take place year 2012 at which time the goal of the CO_2 reduction must be achieved. In this way the devaluation of CO_2 quotas will ensure that individual countries and the world as a whole achieve the target level for CO_2 emission by the year 2012.[3] So, there may be something in the permit market system that could benefit both the US and the EU in future efforts to comply with stringent target levels, such as those imposed by the Kyoto agreement on CO_2 reduction.

Section 2 first analyzes the financial consequences for industry from introducing a tax without refund and compares them with those from introducing a permit market. Section 2 then considers what happens if the tax is refunded to industry. This leads to redistribution within the heterogeneous group of industry and creates labor-intensive winners and energy-intensive losers. Therefore, a permit market is still the politically most attractive solution. Also, the lobbying power of households are analyzed. Section 3 then turns to empirical evidence from Europe and the US Section 4 comes up with a policy recommendation.

Taxation vs. Permit Markets

The regulatory scheme must be politically acceptable which, according to the public choice theory, means that industry (with strong lobby groups) must per-

---

2 The idea of defining and enforcing property rights to common resources stems from Coase (1960) and Dales (1968) and was formalized with respect to permit markets by Montgomery (1972). In contrast to grandfathering, Command-And-Control regulation corresponds to non-transferable quotas as property rights are not transferred to polluters.

3 Countries like Russia, which will reduce and sell CO_2 quotas to the economically more developed industrialized countries, are rewarded financially in this way. Less developed countries will receive important subsidies to make new investments in outdated and run-down industries (Svendsen 1999b).

ceive it as an arrangement consistent with their interests. Therefore, it is relevant to ask what the cost relationship between taxation and a grandfathered permit market is, that is, how much will the regulated parties have to pay in a permit market compared with the private costs associated with taxation without refund of the tax revenue.

Let us consider the case of a Q^* cut in CO_2 emissions. Assume that the marginal costs (*MC*) increase linearly as the reduction of CO_2 emissions (Q) rises, as illustrated in Figure 1 below.

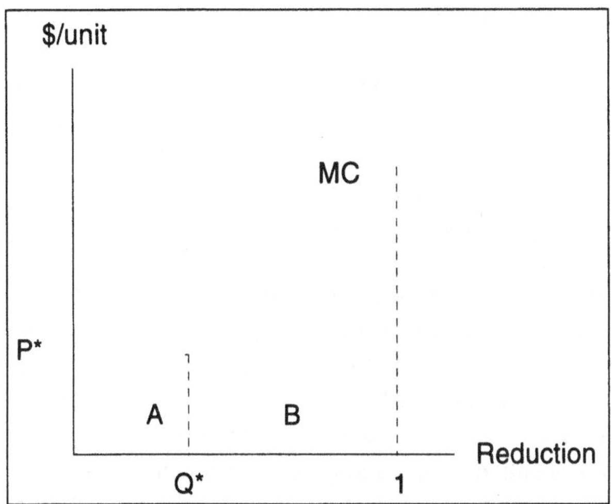

Figure 1: Taxation vs. Grandfathering

Source: Svendsen *et al.* (1999).

Assume, now, that the correct CO_2 tax to accomplish the Q* target reduction level is the price P^*. What are the aggregated costs under a tax and a grandfathered permit market, respectively?

In Figure 1, total costs under a CO_2 tax will be the areas *A* and *B*. Polluters will reduce Q^* of their emissions at the cost to them of area *A*. After that, it is cheaper to pay the tax (than to keep on reducing themselves) on each of the units they continue to emit for a total tax bill of area *B*.

In contrast, the explicit costs to the polluter in a permit market will be *A* only. A tax requires the polluter to pay for all units emitted, whereas under the permit market (at the corresponding permit price P^*), the polluter receives the right to emit the targeted level of emissions for free (according to the principle of grandfathering). Let us compare these costs and divide the area $A + B$ with that of *A*. The triangle and the rectangle areas can easily be calculated by using P^* *and* Q^* (as shown in Figure 1).[4] This gives us:

$$(A+B)/A = 2/Q^* - 1$$

4 $(A+B)/A = B/A + 1 = P^*(1-Q^*)/\ ½P^*Q^* + 1 = (2 - 2Q^*)/Q^* + 1 = 2/Q^* - 1.$

This formula shows the distributional effect when considering tax or permit market at a target level, here denoted by the proportional number, Q^*.

Consider the distributional effects following different target levels. In Denmark for instance, a target level of 20% cut in CO_2 emissions applies (from 1988 to 2005). For $Q^* = 0.2$, we get the result that taxation is nine times more costly to polluters than the grandfathered permit market!

The reduction target level from the UN conference in Kyoto, Japan (December 1997), is also an illustrative example. Here, the industrialized countries voluntarily agreed on reducing CO_2 by 5% from 1990 to 2012. In this case, for $Q^* = 0.05$, it can be calculated that taxation without refund would be 39 times more costly to polluters than the permit market! In the case of the EU CO_2 target level, the negotiated 8% target level from the Kyoto negotiations can also be used as an example. When $Q^* = 0.08$, the permit market is 24 times cheaper to polluters than a corresponding tax without refund. Another example is the US Acid Rain Program, which aims to reduce SO_2 emission by 50%. In this case, for $Q^* = 0.5$, taxation would be three times more costly to the polluters, provided that the MC curve is linear.

Why cannot the tax revenue be refunded in a politically acceptable way? The problem is that the refund must be independent of the pollution. Otherwise, the incentive to reduce pollution would be removed; that is, it would not matter to the polluter how much he discharged or emitted, because all of the money paid in taxes would be refunded. If, for example, a source is given back its CO_2 tax payments, then it will have no incentive at all to reduce CO_2 emissions.

Still, one could argue that increased production costs under the tax solution could be avoided by constructing other types of general refund systems not linked to emission, for example in the form of a reduced tax contribution to labor market services, reduced company tax or reduction in other types of taxes (See Oates 1995 and Stavins & Whitehead 1992).

However, energy-intensive polluters will, as potential losers in a small group, more aggressively oppose taxation with the argument that their competitiveness will weaken; labor-intensive polluters may, as potential winners in a large group, fail to seek a taxation and refund system because they may not mobilize politically since their individual benefits are relatively small and hardly worth the effort. Therefore, the refund system can hardly be modeled such that it would satisfy the small group of potential losers.

This political asymmetry against taxation with full refund may be explained by the size of the group and may be illustrated in the following way. Consider first a group consisting of 1 million identical, small and labor-intensive polluters, where the total value of a tax refund to the group is \$1 million and the total cost of providing it through lobbyism is \$100,000. Further assume that the value of the refund, if provided, would be shared equally among all the members, so each would receive a benefit valued at \$1. This is the case, for example, if each member in the large group pays \$2 in CO_2 tax and receives \$3 in general refund.

Although the group as a whole would get benefits worth $1 million (ten times the $100,000 invested in providing the good), the net benefit to any individual member who chooses to provide the good on his or her own is clearly negative because individual gain is $1 only compared to the individual cost of $100,000! In the absence of political mobilization, the good will therefore not be provided. This group would thus be classified as 'large' and the collective good of tax with full refund will not be provided.[5]

Now consider another group that has only five identical energy-intensive members. This group wants to avoid taxation with full refund because they pay much more in taxes (due to high CO_2 emissions) than they get back from a general refund system. So if this 'small' group loses the same amount that the large group gains, each member will experience a benefit valued at $200,000 from preventing a tax with full refund. This is the case, for example, if each member in the small group pays $400,000 in CO_2 tax but receives only $200,000 in general refund. If the costs of successfully lobbying against taxation is $100,000 again, each individual member's net benefit is $100,000. Therefore, this collective good of avoiding green taxation will now be provided for the small group even in the absence of political mobilization, see Table 1.

Table 1: Individual Net Gain from a General Refund System

	Pro tax (winners) *(labor-intensive)*	**Con tax (losers)** *(energy-intensive)*
Number of firms	1 million	5
Individual tax payments*	$2	$400,000
Total tax payments*	$2 million	$2 million
Individual tax refund	$3	$200,000
Total general refund	$3 million	$1 million
Total gain	$1 million	$1 million
Individual gain	$1	$200,000
Total lobbying cost	$100,000	$100,000
Individual net gain	Negative	$100,000

In this way, the small group of large and energy-intensive firms holds a stronger position than the large group of small and less energy-intensive firms and because

5 Svendsen (1998a). See Olson (1965, 1982) concerning this logic of collective action. See also Paldam & Svendsen (1999) for a discussion on the underlying behavioral assumptions.

of this opposition within the well-organized industry lobby, it is hardly possible to impose a CO_2 tax high enough to achieve the target level in Europe.

Concerning unorganized CO_2 emitters, such as the transportation sector and households, it is reasonable to expect that they will not protest in an organized manner against taxation. This argument follows the logic developed for large groups in this section. The taxpayers are only affected by the tax at the margin, and calculated individual benefits from starting organizing interest group opposition are expected to be smaller than the costs of doing so. Again, it does not pay the individual taxpayer to protest and provide the collective good of avoiding CO_2 taxation.

Experiences from Europe and the US

In practice, green taxes are differentiated and far from the first-best optimal design indeed. For example, empirical findings on CO_2 taxation confirm the theoretical prediction from Section 2. For example, individual member states in the EU have tried to introduce CO_2 taxation on their own. When looking at four EU member states (Finland, Sweden, Denmark and Holland) and one outsider (Norway), Svendsen *et al.* (1999) found that on average, households are met with a tax rate almost four times higher than that of the industry. Further, if the tax advantage of the industry is included, it increases to more than five times the size. This clearly indicates that the empirical findings seem to confirm the theoretical conclusions on organized interest groups influencing the tax policy decisions (Svendsen 1998d).

Also, it has not been politically feasible to introduce a common CO_2 tax in the EU. When the EU Commission launched a $15 tax directive proposal in 1992, this proposal was opposed by several member states. By 1994, the proposed common CO_2 tax in the EU was 'dead'.[6]

In the US, the recent experience with the so-called 'Acid Rain Program' (ARP) deals with a 50% reduction of SO_2 emissions from the 1,000 biggest electric utilities owned by 200 companies (Svendsen 1998c). The overall conclusion is that the ARP has performed well. A competitive market structure and auction mechanism have succeeded in generating extensive trade activity and prices. Property rights to permits have been well-defined and well-enforced and transaction costs are low.

6 Great Britain is against common EU taxes in principle and wishes to deal with the CO_2 problem on its own. Spain demands the right to increase its CO_2 emissions by 25% because its industrial level is lower than that of other EU member states. Portugal and Greece have argued against the CO_2 tax for the reason that taxation could slow economic growth. Spain, Portugal and Greece refuse to accept further production costs until they reach an economic level similar to that of the more wealthy northern EU members (Svendsen 1998b).

The main reason for low transaction costs in the ARP is the fact that this program ignores source location and the risk of creating hot spots by trade. Individual administrative control procedures for each trade were unnecessary because of favorable geographical and meteorological conditions and the 50% SO_2 reduction.

It is important, that one central authority, like the Environmental Protection Agency (EPA) in the US, enforces the market so that local (or national) authorities are not responsible for the control and thus are tempted to protect their own firms. In the case of the EU, the European Environment Agency in Denmark could be an appropriate choice for the enforcement of property rights in a potential CO_2 market (Svendsen 1998c).

The incentive among electric utilities to participate in a potential CO_2 market will grow stronger in the future when competition increases, following the introduction of a single market in the EU.[7] Other firms or sectors may also be interested in the system as well if their energy costs are significant. Therefore, as in the ARP, an 'opt-in' possibility should be added to the program so that these other firms may subsequently apply for participation. Also, an annual reserve for direct sale and for revenue-neutral auctions should be included so to 'kick-start' the market by delivering a price signal to the market. This price signal lowers transaction costs (Svendsen & Christensen 1999, Svendsen 1998d).

Policy Recommendation

The political failure of introducing CO_2 taxation in the EU points to a grandfathered CO_2 permit market for three reasons. First, no revenue needs to be redistributed and, presumably, there will not be political opposition against this policy within any industry as no big losers occur from the regulation. Second, in contrast to taxation, the institutional structure in the EU allows a permit market, as a nonfiscal measure, to be settled on by majority rule. Third, the free-rider problem may be overcome by allocating a relatively larger proportion of the permits to those EU members that are reluctant to participate than their historical emission levels qualify them for. A design similar to that of the US Acid Rain Program (ARP) could be considered whereby the initial group would be EU industry, e.g. all fossil-based electric utilities in the European Union with a production capacity greater than 25 MW (Svendsen 1998b). This market will probably roughly match the size of the ARP but it must be investigated in more detail.

This new policy design is therefore recommendable for future environmental regulation, especially for cases in which source location may be ignored. A first suitable target group for a potential CO_2 market in the EU would be the electric

7 Svendsen (1998b). A similar development concerning third party access has taken place in the US (Svendsen 1995).

utility industry. They should choose for themselves whether or not they want to participate in the market. If they want to participate, they must buy and install a continuous emissions monitoring system (CEMS) monitor. If a source exceeds its quota, it will automatically be fined. The level of the fine could, e.g., be five times the expected market price. Furthermore, the source will have to reduce the excess amount in the following year.

The three main interest groups in the US (private business, environmentalist groups and the electricity sector) prefer a grandfathered permit market. Business is attracted by this solution because free initial distribution of permits both favors existing sources financially and furthermore creates a barrier to entry for new firms. Environmentalist groups have changed attitudes and promote the idea too as a way of negotiating higher target reduction levels with industry. Finally, electric utilities prefer a grandfathered permit market, and this step towards less planned economy may be explained by the rise of competition in the US electricity sector (Svendsen 1999a, 1998d).

It is suggested that green taxes are inappropriate in relation to organized interests such as industry. The tax alternative presents the problem that tax revenue is difficult to refund in a politically acceptable way where these interests are concerned. Energy-intensive firms will lose from taxation even with full refund and are able to protest quickly and with success. This behavior contrasts with that of the latent group of potential net winners, which are typically numerous and small service firms unable to organize.

So even if the tax were refunded in a transparent and general way the political opposition against CO_2 taxation would probably persist because energy-intensive firms, which are also the most energy-intensive, are normally the largest and can organize for collective action – they may behave as a single collective actor or small group. Therefore, the political opposition against CO_2 taxation is likely to stay asymmetrically in favor of potential losers. The overall problem is that potential winners cannot organize their lobbying powers to counteract the potential losers in the political decision-making process.

A permit market, on the other hand, is politically more attractive to the organized polluters than a tax scheme due to the possibility of a free, initial distribution (grandfathering). We derived a formula for the distributional effects of these two economic instruments. It showed that for target levels, such as the Danish case of a 20% CO_2 reduction or the EU case of 8% CO_2 reduction, the tax solution (without refund) was, respectively, 9 and 24 times more costly for polluters than the grandfathered permit market, provided that the MC curves are linear.

These suggestions on how to fill the gap between economics and politics in designing cost-effective and politically attractive instruments, point to the use of a permit market in relation to well-organized interests. In contrast, a CO_2 tax should be applied to non-organized interests in Europe, such as households and the transportation sector. These interests are not well-represented in the political arena because the individual benefits from organizing interest-group opposition are smaller than the added costs of doing so.

References

Baumol, W.J. & Oates, W.E. (1988), *The Theory of Environmental Policy*. 2. ed. New York, Cambridge University Press.

Coase, R. H. (1960), The Problem of Social Cost. *Journal of Law and Economics* **3**, 1-44.

Cropper, M.L. & Oates, W.E. (1992), Environmental Economics: A Survey. *Journal of Economic Literature* **XXX**, 675–740.

Dales, J.H. (1968), *Pollution, Property, and Prices*. Toronto, Ontario, University of Toronto Press.

Green, D.P. & Shapiro, I. (1994), *Pathologies of Rational Choice Theory: A Critique of Applications in Political Science*. Yale University.

Hahn, R.W. & Stavins, R.N. (1992), Economic Incentives for Environmental Protection: Integrating Theory and Practice. *American Economic Review* **82**, 464–68.

Howe, C.W. (1994), Taxes *Versus* Tradable Discharge Permits: A Review in the Light of the U.S. and European Experience. *Environmental and Resource Economics* **4**, 151–69.

Montgomery, W. David (1972), Markets in Licenses and efficient Pollution Control Programs. *Journal of Economic Theory* **5**(3) 395-418.

Oates, W.E. (1995), Green Taxes: Can We Protect the Environment and Improve the Tax System at the Same Time? *Southern Economic Journal* **4**, 915–22.

Olson, M. (1982), *The Rise and Decline of Nations*. New Haven, Yale University Press.

Olson, M. (1965), *The Logic of Collective Action*. Cambridge University Press, Cambridge.

Paldam, M. & Svendsen, G.T. (1999), An Essay on Social Capital: Reflections on a Concept Linking Social Sciences. *European Journal of Political Economy*, at press.

Stavins, R.N. & Whitehead, B.W. (1992), Pollution Charges for Environmental Protection: A Policy Link Between Energy and Environment. *Annual Reviews (Energy and Environment)* **17**, 187–210.

Svendsen, G.T. (1999a), US Interest Groups Prefer Emission Trading: A New Perspective. *Public Choice* **101**(1/2) 109-128.

Svendsen, G.T. (1999b), The Idea of Global CO_2 Trade. *European Environment,* scheduled publication date: Dec. 1999, 9/6.

Svendsen, G.T. (1998a), *Public Choice and Environmental Regulation: Tradable Permit Systems in United States and CO_2 Taxation in Europe*. Edward Elgar, Cheltenham, UK.

Svendsen, G.T. (1998b), Towards a CO_2 Market for the EU: The Case of the Electric Utilities. *European Environment* **8**, 121-28.

Svendsen, G.T. (1998c), The US Acid Rain Program: Design, Performance and Assessment. *Government & Policy* **16**, 723-734.

Svendsen, G.T. (1998d), A General Model for CO_2 Regulation: The Case of Denmark. *Energy Policy* **26**, 33-44.

Svendsen, G.T. & Christensen, J.L. (1999), The US SO_2 Auction: Analysis and generalisation. *Energy Economics* **21**(5) 403-416.

Svendsen, G.T.; Daugbjerg, C.; Hjoellund, L. & Pedersen, A.B. (1999), Consumers, Industrialists and the Political Economy of Green Taxation: CO_2 taxation in OECD. Submitted to *Journal of Public Policy.*

Tietenberg, T.H. (1985), *Emissions Trading: An Exercise in Reforming Pollution Policy.* Washington, D.C., Resources for the Future.

Negotiated Agreements and 'Flexible Mechanisms': Building Blocks for Efficient Kyoto Implementation Strategies in the European Union?

Peter Zapfel [1]

European Commission, Environment Directorate-General,
Rue de la Loi / Wetstraat 200, BU-5 4/125, B-1049 Brussels, Belgium,
e-mail: Peter.Zapfel@cec.eu.int.

Introduction

In December 1997 international climate protection efforts culminated with the historic accord reached at Kyoto. In the Protocol, named after the Japanese city Kyoto hosting the Third Conference of the Parties to the Climate Convention, 38 industrialized and transition countries plus the European Community have committed to limit or reduce emissions of a set of six greenhouse gases[2] (GHG). The European Community and its Member States have the obligation to reduce GHG emissions by 8 % in the target period 2008 to 2012 compared to the emissions level in 1990, the overall reduction of all committed countries amounts to above 5 %.

The Protocol is innovative in terms of the instruments that may be applied in order to achieve the emissions objectives. It authorizes a set of so-called flexible (Kyoto) mechanisms, which may be used in addition to domestic policy efforts

---

1 The author was working for the Directorate-General for Economic and Financial Affairs at the writing of this article and expresses solely his personal opinion.

2 Besides the main greenhouse gas carbon dioxide (CO_2), the Protocol covers also methane (CH_4), nitrous oxide (N_2O), hydrofluorocarbons (HFCs), perfluorocarbons (PFCs), and sulphur hexafluoride (SF_6).

and need to be further developed in order to be operational, and allows participating countries to use them as supplements to whatever other instruments may be considered most effective and deemed politically desirable.

The European Commission (1998a, 1999), as national Member State ministries, has initiated intensive work and a policy debate on the implementation of the Community GHG reduction target. Among others the European Commission (1999) announced its intention to produce a consultation paper[3] in the year 2000 on the potential for emission trading in the European Community, possibly tested in a pilot phase as of 2005.

The European implementation debate to date is characterized by a strong preference expressed by industry stakeholders to use negotiated (long-term) agreements[4] on energy efficiency as the primary instrument to achieve the sector's contribution to the Community GHG objective. At the same time, industry has also expressed an interest in using the flexible mechanisms as a supplement to negotiated agreements. These rather general positions have not been complemented with concrete proposals so far about how such a preferred policy mix could be designed. This paper attempts to facilitate the next step and investigates potential models for combined policies of negotiated agreements and flexible mechanisms.

Section II discusses one of the issues at the heart of the implementation debate with European industry – relative vs. absolute emission targets. Section III and IV develop combinations of negotiated agreements with project-based mechanisms and emission (allowance) trading respectively.[5] In section V we draw conclusions.

Relative or Absolute Targets: A Real Choice?

The kind of commitment which has been developed and agreed by countries in the Kyoto Protocol is a strong form of an absolute target. Based on the estimated emissions level in 1990 (as reported to the UN Climate Secretariat in national inventories) and a reduction / limitation percentage (indicated in Annex B of the Protocol) each country is restricted to a quantitatively limited absolute amount of emissions (an emissions budget called "assigned amount") in the target period 2008 to 2012. These assigned amounts are legally-binding, independent of the development of external factors up to and over the commitment period. Higher than expected economic growth might render the target more challenging than

---

3 Meanwhile, this consultation paper has been released (European Commission 2000).

4 A comprehensive overview, including a definition and numerous examples, of negotiated agreements can be found in OECD (1998).

5 The discussion will be mainly focused on aspects of target setting / fixing the environmental objective and touch only in a limited manner on other crucial aspects for negotiated agreements and flexible mechanisms, e.g. emissions monitoring and reporting, compliance and enforcement provisions.

assumed at the time of Protocol negotiation, while moderate growth may allow a country to save part of the assigned amount and carry it over into the budget of the subsequent target period, or to transfer it to another country during the first target period.

In discussions with some industrial associations on how to determine their respective sector's share in implementing Kyoto, reservations about the use of such absolute emission targets are generally expressed. Industrial associations have pointed out that the traditional type of targets defined as relative improvements in energy efficiency are also their preference in the context of climate policy (e.g., CEFIC 1997a). Absolute GHG targets allocated to a sector would in their view rather constitute an arbitrary "limit on economic growth" and could in the worst case imply a loss of market share for European producers. They reason further that the lost output could be substituted by imports from other countries with lower energy efficiency performance and hence the global atmosphere might even be worse off.

However, from the regulators' / policymakers' point of view such a reluctance by industrial associations to consider absolute GHG targets and commit only to relative energy efficiency improvements creates a variety of problems. Firstly, the Kyoto targets accepted by the European Community and its Member States are of an absolute nature. The regulator is therefore confronted with a considerable degree of uncertainty, if the contributions of some sectors depend upon ex-ante uncertain economic growth patterns. Secondly, the relatively constrained sectors have to face the uncertainty of being targeted on short notice (close to the end of the first commitment period) with inflexible policies, should they experience unexpected strong output and emissions growth. Thirdly, such a stance also implies uncertainties for other societal and economic sectors that do accept absolute targets and stick to it. If a sector with a relative commitment experiences strong growth and substantially higher emissions than reasonably expected, other sectors may have to be asked for additional contributions above and beyond their initial absolute targets in order for the European Community to fulfill the commitment.

The following table assesses quantitatively the potential range of uncertainty introduced by GHG objectives fixed in a relative manner.

Table 1: Relative Targets and Absolute Emissions [%]

Case	Output growth p.a.	Carbon efficiency improvement p.a.	Emissions in 2008 to 2012 (2000 = 100 %)
A1	3.0	2.0	109.8
A2	3.0	1.5	115.9
A3	3.0	1.0	121.6
A4	3.0	0.5	127.9
B1	1.5	2.0	94.8
B2	1.5	1.5	99.8
B3	1.5	1.0	105.0
B4	1.5	0.5	110.4

The figures need to be interpreted as follows: We assume a negotiated agreement implemented as of 2000 with targets determined until the end of the first commitment period in 2012. Two important assumptions are varied: annual output growth, and annual carbon efficiency improvement. The final column indicates the impact of various scenarios of output growth and carbon efficiency developments on absolute emissions in the period 2008 to 2012.

Cases A1 to A4 assume 3 % annual output growth, and vary annual carbon efficiency improvements between 0.5 % and 2.0 %. These cases show very clearly that despite substantial and sustained carbon efficiency improvements over a 12-year period absolute emissions would nevertheless have increased by between 10 % and 28 % on average in the target period. This demonstrates the uncertainty a policymaker faces in the case of a sector committing to relative targets. If we further assume that this sector contributes about 5 % to country-wide emissions, and we regard A1 to A4 as the most likely scenarios, such a negotiated agreement would introduce an uncertainty of about 1 % to the country inventory (since total emissions from the sector would increase between 0.5 % to 1.4 %).

Cases B1 to B4 assume a moderate 1.5 % annual output growth between 2000 and 2012 and the same variations in carbon efficiency. Case B1 shows that even a scenario of moderate growth and an impressive annual carbon efficiency improvement of 2.0 % sustained over twelve years fail to achieve an 8 % reduction of absolute emissions – corresponding to the overall Community target.

The figures leave us with an important intermediate conclusion: a negotiated agreement on carbon efficiency, under realistic assumptions about output growth and efficiency improvements[6], will result most likely in absolute emissions increases rather than decreases, even if a sector commits to substantial advances in carbon efficiency. If output grows stronger than carbon efficiency improves, efficiency-based agreements will always result in absolute emissions increases.

Therefore, and if one does not want to impede output growth, one would have to go substantially beyond efficiency improvements recorded in the past and ask: what ratio of sustained carbon efficiency improvement is necessary to achieve a given absolute emissions level under the assumption of a certain output growth rate?

Table 2: Efficiency Improvements under Alternative Absolute Targets [%]

Targeted emissions level in 2008-2012 against 2000	Output growth p.a.	Sustained annual carbon efficiency improvement
92	+ 3.0	3.72
92	+ 2.0	2.78
92	+ 1.0	1.81
100	+ 3.0	3.00
100	+ 2.0	2.00
100	+ 1.0	1.00
84	+ 3.0	4.59
84	+ 2.0	3.66
84	+ 1.0	2.70

The above Table 2 demonstrates that an 8 % reduction – the target for the European Community, assuming that emissions in the year 2000 are stabilized at the 1990 level – necessitates a high level of sustained carbon efficiency improvement over more than a decade even under moderate output growth. An extended stabilization would require efficiency improvements to offset emissions growth caused by output increases, while a 16 % reduction would constitute a need for even greater efficiency improvements.

6 To give an indication of the realism of the used figures we have looked at production growth in the European chemical industry (CEFIC 1997b), one of the most interested sectors in negotiated agreements. This sector has increased output by an annual 3.4 % over the period 1985 to 1996 and 3.1 % between 1990 and 1996. Carbon efficiency in the European Community (European Commission 1998b) has improved by about 2.4 % annually between 1985 and 1990. Up to 1994 the annual improvement has slowed down to 1.8 %. Since then it is further declining. Finally, energy efficiency has improved by about 1.6 % p.a. in the eighties and 0.6 % p.a. in the early nineties.

We see that the reluctance of some economic sectors to commit to absolute GHG emission targets introduces a considerable degree of uncertainty for the regulator. The presented figures suggest that any consistent and reliable Kyoto implementation strategy should be based on absolute GHG emission targets agreed with all major emitting sectors, either in an explicit or implicit manner. While explicit absolute targets are self-explanatory, we understand as implicitly assumed absolute targets a planning assumption used and sufficiently publicized by the regulator. The implicit target could also be determined, if all but one (n-1) sectors agree to absolute targets as the remaining part of the total assigned amount.

Hence one of the primary challenges for the policy designer is how to achieve (at least implicit) quantity certainty in an environment characterized by relative targets agreed with some important sectors. In the following two sections we explore the possibilities of supplementing negotiated agreements with flexible mechanisms. One of the objectives is to devise models which guarantee a sufficient degree of ex-ante quantity (emissions) certainty for the policy designer. This assumes that if the affected industry risks overshooting the – explicitly or implicitly – agreed or assumed absolute emissions target for whatever reason (high output growth and / or low efficiency improvements), the sector might be able to avoid other regulation by purchasing emission reductions, i.e. participating in the flexible mechanisms market.

Negotiated Agreements and Project-Based Mechanisms

For the subsequent analysis we assume the sector to be constrained by a relative carbon efficiency target fixed in a negotiated agreement and to desire the option to use project-based emission reduction credits to contribute to the fulfillment of the implicit absolute commitment.

The discussion is based on the following example:

> A sector produces 100 output units and generates 10 units of carbon emissions at the outset. In a negotiated agreement the sector commits to a 20% carbon efficiency improvement. Hence, the 100 product units 'shall' generate only 8 units of carbon emissions. These 8 units could serve as the implicit target assumed by the regulator in planning for the sector's contribution towards the overall target.

Two uncertainties enter the equation to determine the absolute emissions level resulting from a negotiated agreement on carbon efficiency. Firstly, the sector may fail to fully achieve the efficiency improvement. Moreover, because of growing product demand the sector may increase its output (or increase it by more than assumed under the implicit target). The uncertainties run though in both directions and could as much result in an over- as an under-achievement of the implicit absolute target. We look at the two cases in the following.

Case 1: Under-achievement of the implicit absolute target

> The sector falls short and achieves only 15 % carbon efficiency improvement and its product output grows by 10 % in the regarded period. So in the end the sector will produce 110 units of output and emit 9.35 units of carbon emissions. Of the 1.35 units increase in absolute emissions above and beyond the value expected by the regulator, 0.55 units are due to insufficient carbon efficiency improvements, while 0.80 units are caused by increased output.[7]

The regulator has two basic alternatives to react to this emissions overshoot. On the one hand, he could introduce some regulation (e.g., technical control measures) in order to bring the sector's emissions down to the necessary level. Alternatively, the efficiency-based negotiated agreement could be supplemented by project-based mechanisms in the way that the sector could make use of project-based (joint implementation, JI, or clean development mechanism, CDM) reduction credits in order to make up for

1. the part which could not be realized by efficiency improvements within the sector (0.55 units in the above example)
2. or the part of absolute emissions growth due to output growth (0.80 units)
3. or the total overshoot due to both insufficient efficiency improvement and output growth (1.35 units).

The opening of negotiated agreements in one or the other (1 or 2) or in both (3) directions offers the possibility to tap into cheaper reductions outside the sector and may induce sectors faced with relatively high marginal abatement costs to commit to more ambitious GHG emission reductions in contributing a share to overall efforts.

Case 2: Over-achievement of the implicit absolute target

> The sector achieves 25 % carbon efficiency improvement and its product output declines by 5 % in the regarded period. So in the end the sector will produce 95 units of output and emit 7.125 units of carbon emissions. Of the 0.875 units decrease in absolute emissions below the value expected by the regulator, 0.475 units are due to additional carbon efficiency improvements, while 0.4 units are caused by decreased output.

In such a case of over-achievement and under the assumption that all other sectors in the respective country fulfill their commitments, the country itself would

7 In strict mathematical terms 0.05 of the 1.35 units of the absolute emissions deviation can not be contributed exclusively to either output increases or failed efficiency improvements, but are the result of a joint effect. We disregard this analytical complexity in the paper and attribute the joint effect for simplification to the efficiency side.

over-achieve the absolute Kyoto target. If the sector has only committed to an efficiency target, it cannot directly profit from lower than expected output growth (or abatement costs), or higher than expected efficiency improvements, as would be possible in a situation in which the implicitly assumed absolute target is accepted by the sector and made explicit through the allocation of permits. Hence the country as a whole (e.g. by banking or selling a part of the assigned amount) and / or other economic sectors (e.g., by emitting beyond the absolute target allocated to them) may profit, but no reward could be earned by the sector itself.

One may consider as a possible remedy to allow the sector a reward by hosting a JI project and selling the reductions as project-based credits. However, such a solution would suffer from a number of drawbacks that would make it less attractive than a combination of a negotiated agreement with full-fledged emissions (allowance) trading, as discussed in the subsequent section of the paper. Emission reductions generated by hosting a JI project would require applying the (not yet determined) administrative procedures of such projects (e.g. determination of baselines), which are very cumbersome and may cause quite high transaction costs. Emission reduction credits generated from lower output growth or higher efficiency improvements could only be determined and claimed at the end of a compliance period (ex-post) with a certain time lag necessary for administrative clearance. At that point in time, other economic sectors would have executed their compliance strategies and there might no longer be a demand for the reduction credits. Even more fundamentally, it is doubtful whether reductions against an implicit target – serving only as a working hypothesis for the regulator – will pass the additionality requirement central to JI projects according to Article 6 Para. 1(b) of the Kyoto Protocol.

It is unlikely that such a combined policy approach would result in cost-effective abatement of greenhouse gases. The condition for cost-effectiveness is equalization of marginal abatement costs of emission reductions across all firms and sectors. Since the sector is only under certain circumstances – and with high transaction costs – exchanging credits with other sectors, cost-effectiveness across sectors appears rather unlikely. Cost-effectiveness within the sector, across the affected companies, depends crucially on how the overall efficiency improvement is shared out among firms. Again, there is no a priori reason to expect that cost-effectiveness will be attained.

Negotiated Agreements and Emissions (Allowance) Trading

Emissions (allowance) trading – at the sectoral, domestic, or international level – requires as an indispensable condition the setting of absolute targets (caps) and (ex-ante) allocation of emission allowances to participants. For this reason such a system is frequently called "cap-and-trade".

The only viable option to combine negotiated agreements with emissions (allowance) trading presupposes the acceptance of absolute targets by the sector. In fact, such an agreement would mainly contain the quantity of the assigned amount allocated to the sector (the target), and an industry declaration on how it intends to reach this target. Hence we base the further discussion on the conclusion of such an agreement with absolute targets.

The key to the combination of a negotiated agreement with emission trading is then to use the consensus-based process central to the development of a negotiated agreement for target setting (fixing the environmental objective), while allowing for unfettered emission trading in the implementation of the objective.

Such a system could be opened for exchange of permits with other sectors – nationally or internationally – that are constrained by absolute targets and have been allocated permits. Cross-sectoral / inter-industry trading, based upon sound emissions monitoring, compliance and enforcement provisions, appears both from the perspectives of environmental effectiveness and economic efficiency as a promising choice. A functioning market would guarantee a cost-effective outcome across all participating sectors and firms. And the absolute emission caps, supplemented by accurate emissions monitoring and stringent non-compliance rules, will secure the environmental outcome.

In view of the political difficulties that may arise from committing to an absolute GHG target for some sectors interested in working with this combined approach, one could imagine models that do not foresee an absolute target introduced ex-ante, but a 'de facto' absolute target determined ex-post. A possible model is to work with a relative target, but fix the target only after the output figures are known:

> A sector is supposed to emit no more than 10,000 tons of carbon per year in order to allow for compliance of the country with the absolute Kyoto commitment. In the year 2008 the sector produces 25,000 units of output. Once this total output figure is known a relative target of 0.4 tons of carbon / unit of output is fixed and 2008 allowances are allocated ex-post to all firms in the sector based on actual production figures and the efficiency target.

A major drawback of such a model is the late distribution of allowances. Hence companies gain certainty about the exact number of allowances allocated and can start exchanging them only at a late stage. A way to overcome this problem would be to tentatively allocate e.g. 90 % of allowances as based on the production figures and relative target of the preceding year (hence about 9 of 10 allowances are allocated ex-ante). Once production figures for the year are collected, the efficiency target is fixed and the allocation adjusted. Companies either receive additional or surrender excessively allocated allowances. However, such a system of output-based allocation is seen as problematic from the theoretical point of view (for details see Fischer 1997).

A model of negotiated agreements and emissions (allowance) trading raises a set of technical issues, which we have disregarded in the rather focused discussion so far. These issues are not examined conclusively but rather raised in order to point out their importance in a further treatment of the topic.

One obvious issue in the European Union (EU) context is whether a combined policy of negotiated agreements and emissions (allowance) trading should be introduced at the EU or Member State level. The response to this question depends among others on the structure of the sector with more integrated sectors necessitating rather a European-wide scheme, while firms competing in sectors with parallel, less integrated national or local markets may also be tackled with Member State schemes. In the end, the economic outcome of a system introduced at the EU level or of 15 integrated Member State schemes may be identical.[8] However, major differences may arise as regards the administrative efforts and costs of e.g. the introduction of legislation, operation and maintenance of separate national permit registries, compliance and enforcement regimes[9] compared with a Community-wide scheme. This would suggest at a minimum that design of Member State schemes should be based upon "Community guidelines", both to avoid duplication of administrative costs and simplify the integration of Member State permit markets.

A second issue concerns the main actors in the permit market. Should a scheme be developed – i.e., targets agreed and permits allocated – with the sector (industry) association or with the firms that make up the sector (the members of the association)? Allocating permits to the association would delegate the difficult decisions of how many permits to allocate to the individual firm or facility up to consensus among the firms that make up the sector. A number of legal issues would have to be tackled in the development of such a policy. For example, how would an association – if it is allocated the permits in the first instance – make sure that companies it represents do not leave the association during the period of implementation of the negotiated agreement? In such a case the association could no longer live up to the commitment made to cap emissions in its sector and environmental effectiveness may be jeopardized. There may be a need for a double-

---

8 The primary requirement for cost-effectiveness is the equalization of marginal abatement costs across all market participants. While a Community-wide scheme would guarantee such an outcome, 15 parallel Member State permit schemes would most likely result in 15 differing scarcity values (permit prices). However, if one allows trading among market participants in different Member States (i.e. integration of the individual Member State schemes in an internal market), assuming a functioning market, the outcome would be equivalent to a Community-wide scheme in permit price, total abatement costs, chosen abatement activities etc.

9 A possible consequence of integrated Member State permit markets with differing financial penalties for non-compliance could be that a company operating and participating in allowance markets in more than one Member State would intentionally choose to be in non-compliance, if they run into compliance problems at all, in the Member State with the lowest financial penalty.

layered compliance and enforcement structure regarding emissions monitoring and reporting as well as permit holding and surrendering. The firms would have to surrender permits to the association or be sanctioned, if permits fell short of actual emissions. And the association would have to surrender permits to the regulator or be subject to sanctions.

A final issue is the market structure of the sector for which a policy mix of agreements and flexible mechanisms is developed. Not only does the structure of a sector have an impact on the choice of allocating permits to the sector association or to individual firms, the output market structure may render the combined policy more or less attractive. E.g. a sector with an atomistic market structure (many and very small competitors) may render it impossible to develop a policy mix with an association, because of incomplete representation due to free-riding behavior.[10] For the same reason that an association cannot represent all the players in the sector, it may not be possible for the regulator to work out a scheme directly with all the firms around a table and alternative policy instruments (e.g. carbon or energy taxation, limit values) may prove more effective. A market structure (in the underlying product market) with a large dominant player and many small competitors may not prove useful for intra-sector emission trading either, since it could allow the dominant firm in the sector to exert additional pressure on the smaller actors constituting the competitive fringe by means of hoarding or other price-influencing behavior in the permit market.

Conclusions

This paper has investigated potential models for the combination of negotiated agreements with flexible mechanisms. One of the main objectives of the exercise was the identification of policy mixes of negotiated agreements and various forms of flexible mechanisms which allow for a sufficient degree of ex-ante absolute emissions quantity certainty for the regulator. A number of conclusions emerge from the analysis.

Firstly, we find that the Kyoto climate challenge requires a move beyond the prevailing paradigm of relative policy variables – energy efficiency and carbon intensity – to policies that target absolute emissions performance in a more direct manner.

In order to develop a coherent implementation strategy that minimizes risks of non-compliance with the Kyoto targets and avoids or minimizes 'late into the day' government interventions of an inflexible regulatory nature, the analysis points to

---

10 It is worthwhile to note that some Member States (e.g., Austria) with systems of mandatory membership in the employer federation(s) may be less prone to this problem than others with voluntary associations.

the need to opt for a policy mix that provides the regulator with an acceptable degree of ex-ante certainty about absolute GHG emissions output – at least in an implicit, if not an explicit manner.

We have analyzed several models for combining negotiated agreements with flexible mechanisms in the industrial sector. Our analysis suggests that it is not possible to 'square the circle' and develop a policy mix that provides for a high degree of ex-ante quantity certainty for the regulator, while avoiding absolute emissions limitations for affected industries.

More specifically, we demonstrate that the lack of an explicit absolute environmental target would render a combined scheme of efficiency-based negotiated agreements with project-based mechanisms unlikely to be cost-effective in a cross-sectoral view, if not also from an intra-sectoral perspective. Moreover, it would not allow the affected industry to benefit in a straightforward manner from target 'over-achievement'.

A negotiated agreement based on (ex-ante) absolute targets as a facilitating step towards allowance allocation, combined with emissions (allowance) trading would make a cost-effective outcome more likely, if a number of intricate issues can be solved satisfactorily in the design process.

References

European Chemical Industry Council – CEFIC (1997a), *Voluntary energy efficiency programme – VEEP 2005*. Brussels, November 1997.

European Chemical Industry Council – CEFIC (1997b), *Facts & Figures: The European chemical industry in a worldwide perspective*. Brussels, November 1997.

European Commission (2000), *Green Paper on greenhouse gas emissions trading within the European Union*. COM (2000)87.

European Commission (1999), *Preparing for Implementation of the Kyoto Protocol*. COM (99)230.

European Commission (1998a), *Climate Change – Towards an EU post-Kyoto strategy*. COM (98)353.

European Commission (1998b), *Energy in Europe: 1998 – Annual Energy Review, Special Issue*. Directorate-General for Energy, December.

Fischer, Carolyn (1997), *An Economic Analysis of Output-Based Allocation of Emissions Allowances*. Draft. Resources for the Future, Washington, November.

Organisation for Economic Co-operation and Development – OECD (1998), *Voluntary Approaches for Environmental Protection in the European Union*. Environment Directorate, Document ENV/EPOC/GEEI(98)29/FINAL.

Greenhouse Gas Emissions Trading in a Private Company – BP Amoco's Flexible Mechanisms for an Efficient Climate Policy

Peter Knoedel

Deutsche BP Aktiengesellschaft, Member of the Board of Management,
P.O. Box 600340, D-22291 Hamburg, Germany, e-mail: knoedel@bp.com.

Introduction

The previous contributions tackled some aspects of flexible instruments: what the Kyoto Protocol says about them, how to shape abatement strategies, and what experiences were made with emissions trading in the United States.

This paper will cover:

- BP Amoco's commitments to climate change,
- the Group's emission profile and how we intend to meet our greenhouse gas (GHG) target,
- Pilot Emissions Trading program,
- taking trading Group-wide.

An aspect which I will be trying to add is how my company, BP Amoco, has set out to *put emissions trading to practical use* as one tool to deliver our commitments to address climate change and, particularly, to reduce our GHG emissions. I will do this in four steps, starting by setting some context.

The Company

BP Amoco is one of the three largest private oil companies in the world, and to illustrate this, some key data are listed below (see Figure 1). The announced merger with Arco, the sixth largest US oil company – still under review by US ánd European Union competition authorities – will make us the second largest, in particular by adding to our oil and gas production as well as proven reserves. Our solar energy business, BP Solarex, is the one of the two world leaders in photovoltaics with a global market share of some 20%.

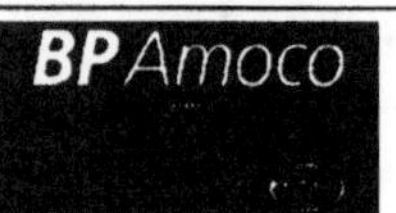

BP Amoco - 1998

- ➔**Market value: 170 bn $ = GDP of Norway or Turkey**
- ➔**Sales 94.5 bn $ = GDP of Portugal**
- ➔**Income 1998: 4.5 bn $; 96,000 employees**
- ➔**Oil production: 2.0 mbd; Product sales 4.4 mbd = 1 1/2 of total German oil consumption**
- ➔**Oil and Gas resources: 15 bn boe = 16 years of German oil consumption**
- ➔**18 own or JV refineries; 27,000 retail stations**
- ➔**BP Solarex: 150 m $ sales**

Figure 1: BP Amoco 1998

In May 1997, John Browne, the chief executive officer (CEO) of BP Amoco, made a presentation at Stanford University in California entitled 'Climate Change: the New Agenda'. The notable speech outlined BP's approach to climate change.

The science of climate change is uncertain and may always be, but that does not exclude the possibility of taking *precautionary* steps.

The production and use of hydrocarbons being one source of anthropogenic GHG emissions, we believe that the oil industry has the ability and the responsibility to contribute to the debate and the solution.

What BP Amoco has proposed to do, regarding the threat of climate change, is substantial, real and measurable. There are considerable risks in doing either too much, or too little, at the wrong moment. Whatever is done must be equitable without significant disadvantage to any group of nations or industries.

BP Amoco believe that the Kyoto mechanisms, including emissions trading, joint implementation and the clean development mechanism, have the potential to provide the appropriate flexibility to deliver cost-effective reductions.

Why is Climate Change Important to Our Business?

As the climate change agenda rises, national governments are planning their implementation of the commitments they made at Kyoto. In Germany, industry has committed to voluntary carbon dioxide (CO_2) reduction targets as a more cost-effective way than eco-taxes or regulatory initiatives. But as these commitments are valid for industry sectors, we will be facing a debate on how to implement this at the level of individual companies.

This is one reason why we feel that climate change may be increasingly seen as a company competitive issue that may differentiate companies in the future.

It is very likely that the issue of climate change will increase in importance in many government policies.

International emissions trading across national boundaries, however, is still some years away. Considerable progress is required by governments to settle guidelines and rules for emissions trading. The Kyoto Protocol allows trading between countries after 2008 with – possibly – trading of credits from clean development mechanism projects (CDM) from 2000. Domestic regulation – whether taxation, efficiency standards or voluntary agreements etc. – is likely to be different in different countries. Several governments are looking at domestic emissions trading schemes – Norway, the United Kingdom, Australia, New Zealand, Canada and others are proposing systems which may come into effect around 2003-2008.

So to be prepared to the extent of having tried and tested some of the Kyoto instruments seems a prudent way of running our business.

BP Amoco has committed publicly to deliver:

- 10% reduction in GHGs of 1990 levels by 2010. This is on a direct emissions basis and also based on the equity BP Amoco hold in operations. That is a firm figure to build into year by year performance contracts of all BP Amoco managers because we see performance as indivisible. There is no trade-off for any manager, at any level, between our environmental targets and our financial targets.
- BP Amoco intends to expand its currently pilot emissions trading system to cover all operations during 2000.

BP Amoco climate change work plan incorporates a portfolio approach to climate change and covers all aspects shown in Figure 2.

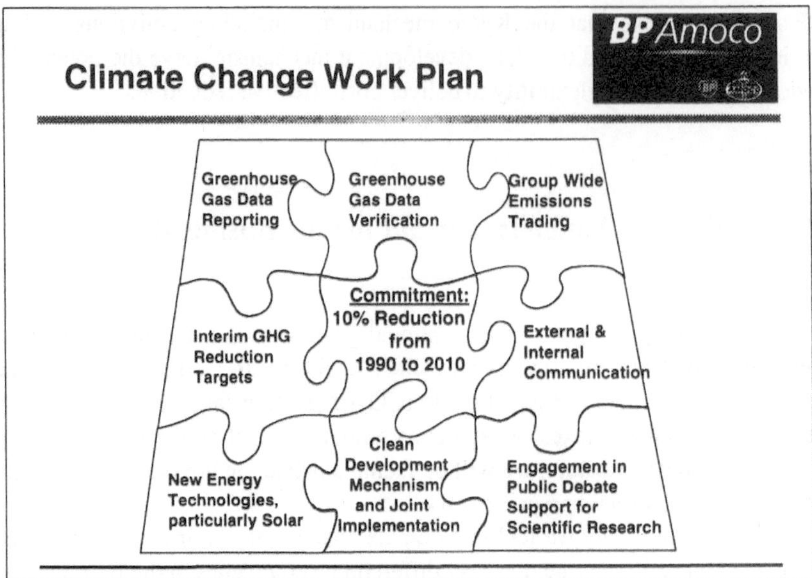

Figure 2: Climate Change Work Plan

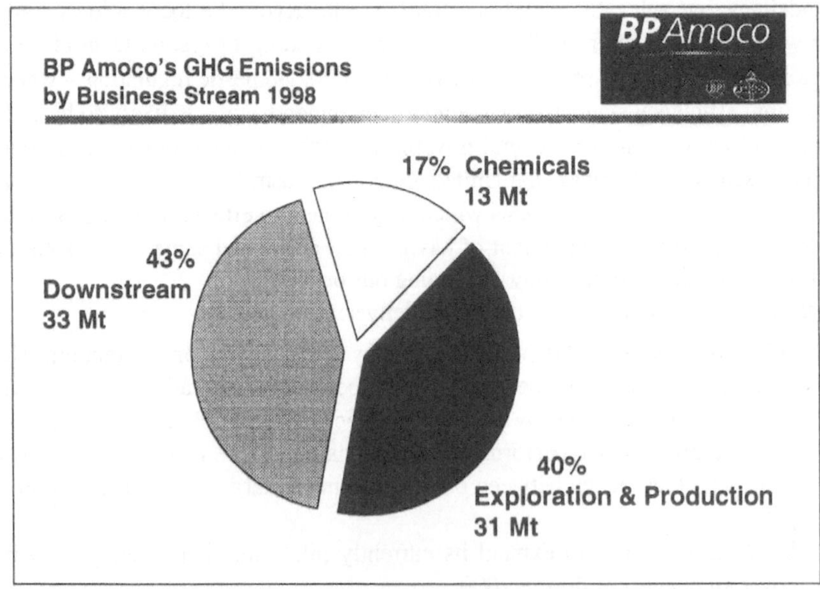

Figure 3: BP Amoco's GHG Emissions by Business Stream 1998

The current best estimation of BP Amoco's GHG emissions from 1998 indicates contributions by Exploration and Production at a similar level as Downstream (refining), with Chemicals the smallest portion (see Figure 3).

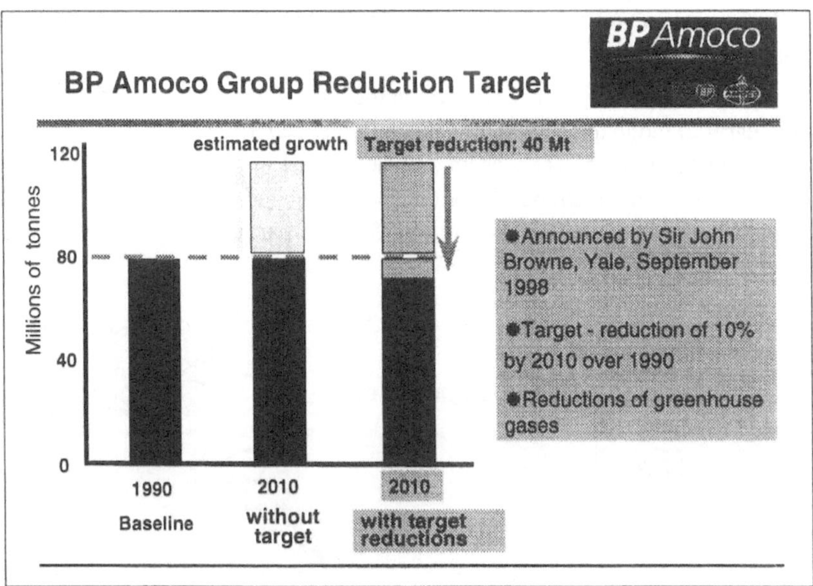

Figure 4: BP Amoco Group Reduction Target

Figure 4 indicates the extent of BP Amoco commitment to reduce GHGs. When considering the business-as-usual growth in emissions, the target in real terms is more like 35-40%.

There are a number of strategic and practical reasons why BP Amoco as a company has embarked upon emissions trading as a mechanism for reducing GHGs:

- Demonstrate commitment to climate change.
- Learn lessons to share externally. From the start of our pilot emissions trading system, BP Amoco has always intended to share experiences and ideas externally – that's one reason for this contribution.
- Develop a cost-effective mechanism for reduction of GHG emissions. As a competitive, commercial business, of course BP Amoco has to be conscious of cost. Therefore, BP Amoco will look for the lowest-cost way of meeting our GHG target. Theory tells us that emissions trading may provide the means of doing so.
- Gained experience in emissions trading will potentially provide us with the competitive advantage for the group as domestic or international systems develop in the future.
- Price discovery of our internal cost of abatement is important considering the extent of the target as was described.

- Stimulating innovation is a crucial element of emissions trading. Already we have noticed a rise on the number of requests from our commercial teams and engineers wanting to understand how they can factor emissions reductions into their project proposals. Perhaps employees are beginning to view GHG in a different way; a way which could add value to their business unit and the company as a whole.

The BP Amoco pilot system went live on the 14th of September, 1998. Although it is too early to judge the results, this is a progress report on the development and the design of the system. The system was developed in partnership with the US-based Environmental Defense Fund (EDF) which had developed considerable expertise in emissions trading (see Figure 5). The partnership has proved to be positive and successful.

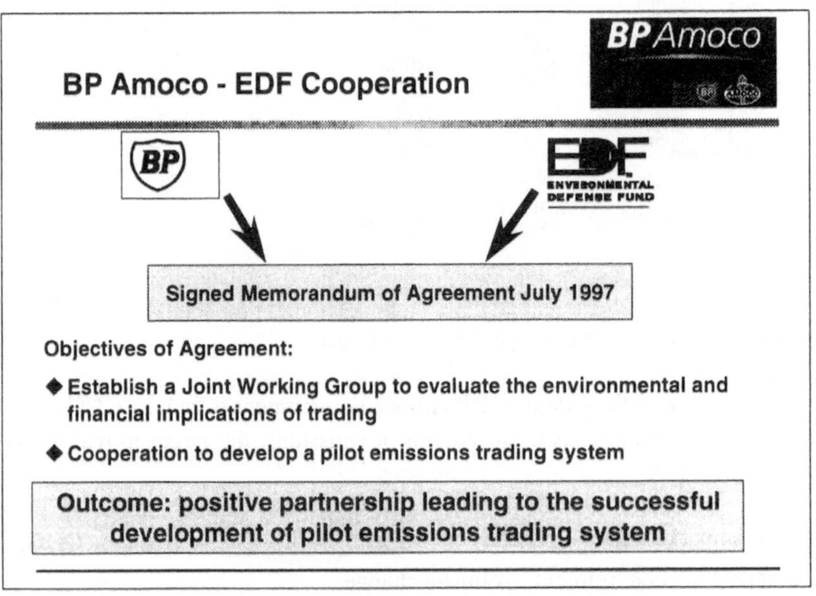

Figure 5: BP Amoco – EDF Cooperation

There are basically two types of emissions trading systems – *allowance trading*, where each player has a defined cap on emissions and then can either buy or sell depending on their future emissions needs and the relative cost of abatement, or *credit trading*. Credit trading is similar to the proposals for joint implementation (JI) and the CDM in the Kyoto Protocol. Credits are generated by evaluating individual projects to estimate reductions compared to a business-as-usual case.

It was decided that *allowance trading* would be the centerpiece of the BP Amoco scheme and that, in the initial phase of the pilot, little use would be made of externally generated credits. In order to put such a scheme in place, two key elements are needed: a target which sets the cap on emissions, or the emissions objective, and a basic allocation of permits to the participating Business Units (BUs) (see Figure 6).

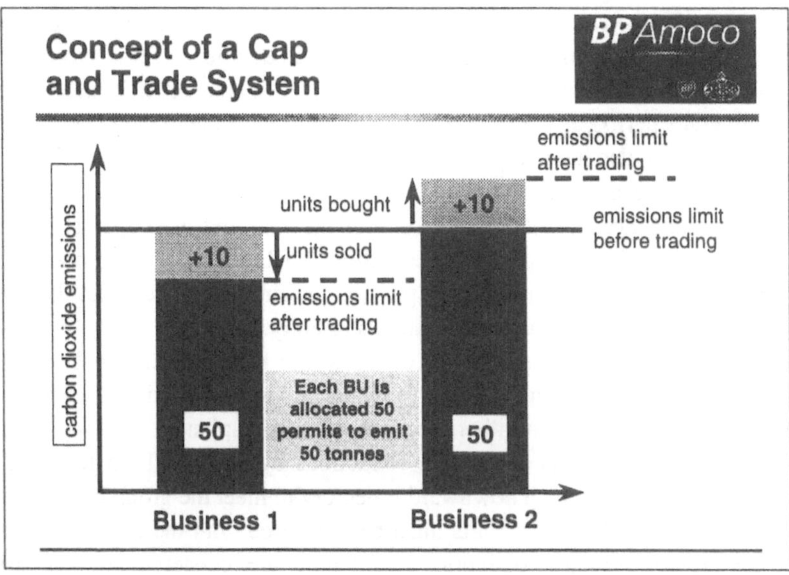

Figure 6: Concept of a Cap and Trade System

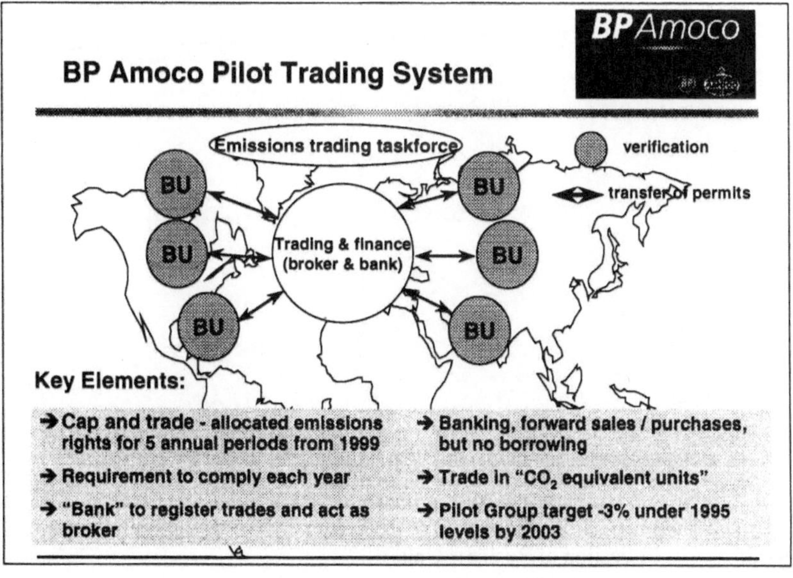

Figure 7: BP Amoco Pilot Trading System

Each participating Business Unit receives an allocation of permits for each of the next five years starting in 1999. Its obligation is to live within that allocation plus any extra permits bought, minus any sold.

At the end of any year, each Business Unit has a sixty day period to make sure that it has sufficient permits to cover its emissions in that year. If it is short, it will need to go into the market as a distressed buyer. If it fails to cover its needs at the end of the grace period, it will face a fine at a multiple of the highest permit price in the year. It will also have to buy permits to cover its shortfall to ensure that the environmental objective is delivered.

Unused permits can be banked at the end of the year and used against a future year's obligation. That gives greater flexibility to a Business Unit to manage its own emissions profile. Banking should promote incentives to overshoot the target early on to build up a buffer against future uncertainty.

These provisions mirror the structure of the Kyoto Protocol exactly. The Protocol does not allow borrowing permits from future years and this is not allowed under the BP Amoco system either, because of the compliance issues that arise. The pilot group target is to deliver 3% reductions by 2003 of 1995 levels. This would set the pilot group on a downward trajectory to meet the group target.

All trades between Business Units must be registered with the broker. Bilateral trades are allowed as long as both price and quantity are registered, but it is anticipated that most trades will go through the broker.

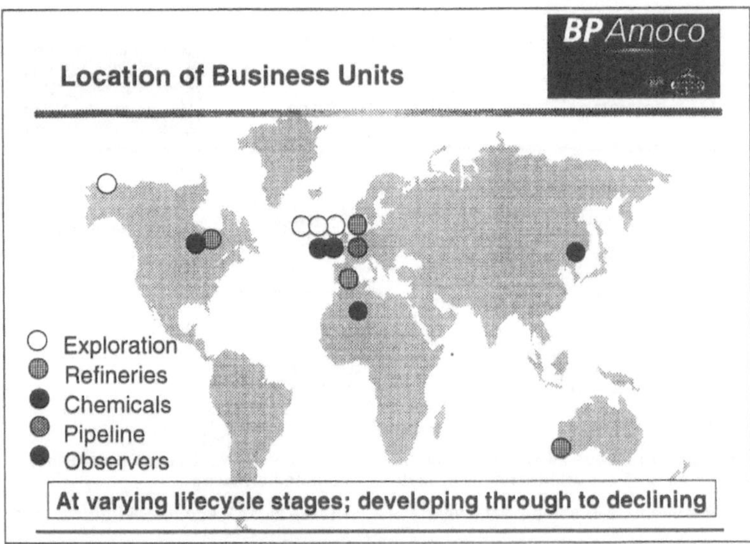

Figure 8: Location of Business Units

Business Unit volunteers were sought from across the three main Business Streams – Exploration and Production, Downstream and Chemicals. Because of the need to make a viable market, only BUs with significant emissions were selected – therefore, all BP Amoco's marketing activities are excluded.

Initially, participating BUs span three continents – America, Europe and Australia (see Figure 8). The oil fields are at different stages in their life cycle and the scale of emissions from BUs ranges significantly. Two non-Annex 1 start-up BUs have been involved in the design work *as observers* but will not be involved in trading in the foreseeable future. Together, these Business Units account for around *10 per cent of BP Amoco's total direct emissions on an equity share basis.*

What Has Happened on the Market?

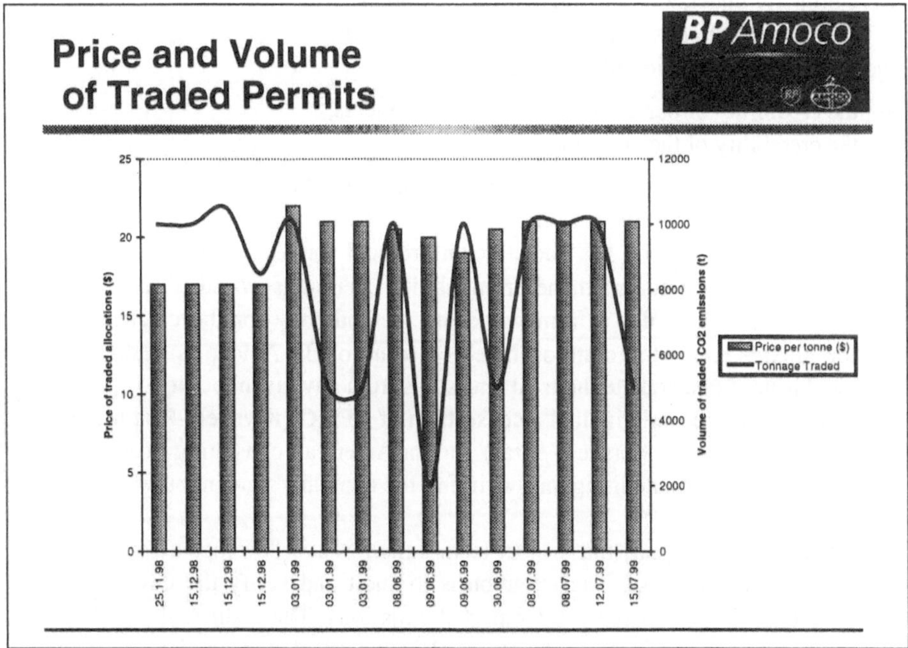

Figure 9: Price and Volume of Traded Permits

The pilot emissions trading system opened for trading September 1998. Since the launch, 15 trades have been carried out representing 121,000 tons of CO_2. Trades have been carried out within the United Kingdom and with the US.

Trading has been limited, part of which is attributable to the huge BP Amoco merger and the changes in structure and dedicated personnel.

What Have We Learnt So Far?

It is difficult to comprehensively evaluate the pilot emissions trading system, due to the limited time period of the system. However, a number of key issues have arisen; these include:

- Raising the awareness of climate change issues throughout the Group.
- Created innovative business strategies to find cost-effective solutions.
- Perhaps most importantly, trading enabled us to financially quantify the GHG implications of investment decisions and – hence – to give "value" to the environment.

There are also a number of technical learning issues which will be incorporated in the design features of the Group-wide emissions trading system, as soon as we have properly understood the impacts on the trading dynamic.

The reporting of emissions is crucial to the credibility of any trading system. We want external verification of GHG data to ensure:

- the credibility of our climate change commitments,
- the credibility of the emissions trading scheme.

BUs, companies and governments must all have confidence in the reporting in order to maintain a credible system.

Since 1997, BP Amoco has had its own protocol for reporting CO_2 emissions and, more, recently for methane from all its operations world-wide. The BP Amoco Protocol provides a framework and methodology for the calculation of carbon dioxide emitted into atmosphere as a result of BP Amoco operations. The Protocol is used to calculate the total emissions from any given facility.

BP Amoco has recently invited a consortium of KPMG (Klynveld Peat Marwick Goerdeler), DNV (De Norske Veritas) and an American consulting group called ICF to begin work on auditing and verifying the company's greenhouse gas emissions.

By the end of the year, the project team aims to have developed a robust and transparent audit process. The intention is to audit and verify the Group's 1990 baseline data as well as the 1998 reported emissions. The audit will also be designed to address verification of data used in the BP Amoco emissions trading system in order to go Group-wide in year 2000.

The project starts with a review of the protocols used by the Business Units to gather and report GHG data. The consortium will then begin to develop the audit and verification process, bringing together their experience in audit methodologies, an awareness of international developments on climate change as well as an understanding of the Group's business operations around the world.

As previously mentioned, BP Amoco intends to expand its trading system.

We are entering new territory and there are still difficulties to be resolved. However, having set up and started the Pilot Emissions Trading System we discovered that many of these issues can be overcome through a process of learning by doing.

Recently, we have been analyzing the GHG data from all our 126 BUs worldwide. This has been useful in helping us to address the all important allocation of emissions rights.

If emissions trading is going to work, BUs must see its usefulness and its advantages over other approaches. It should not be cumbersome in regulation but it must provide a clear framework that gives the BUs an accurate price signal of the greenhouse gas implications of their investment decisions.

We are currently under a consultation period with our BUs throughout BP Amoco to obtain their critical feedback on proposed options. It is our intention in 1999 to work closely with BUs in the development of a system that they feel can achieve this objective.

We must also follow developments of domestic trading systems to ensure appropriate alignment. As our system increases in size and covers a wider array of national boundaries it will come under increasing scrutiny from the external world. Therefore, it is vital that we have confidence in the trading system – that it maintains environmental integrity and helps to achieve GHG reductions across our businesses cost effectively.

We intend to continue our collaboration with the Environmental Defense Fund.

What Conclusions Should Be Drawn from BP Amoco's Emissions Trading Scheme?

There are a number of *generic* key issues which should be *addressed by designers of emission trading systems* in order to deliver GHG reductions at minimum economic cost:

- Clearly, transaction cost of trades should be kept to a minimum.
- Targets and reductions should be devolved upon companies in order to carry out the transactions.
- Credits from joint implementation and the CDM should be equally transferable in trading systems.
- Trading should not be overcomplicated, as this will only discourage trading.
- Finally and perhaps most importantly, companies should be recognized from action to make reductions which are carried out in advance. This is an important incentive for companies such as ours to start making progress immediately.

Meeting the Climate Challenge in the Near Term: Policy Incentives for Early Action

Annie Petsonk and Joseph Goffman

Environmental Defense Fund, 1875 Connecticut Avenue, N.W., Washington, DC 20009, USA, e-mail: annie@edf.org and joseph_goffman@edf.org.

Introduction

If the nations of the world are to meet the goal of preventing dangerous anthropogenic interference in the world's climate system, governmental policy-makers will need to move promptly to develop and implement effective policies to limit the emissions of the greenhouse gases that are contributing to the most rapid warming that planet Earth has experienced in the past 10,000 years. Formulating effective climate change policy, however, presents myriad challenges. This timely conference, organized by the Centre for European Economic Research and counterpart organizations, provides a wonderful opportunity for exchange of views on these challenges and ways of meeting them. The Environmental Defense Fund very much appreciates the invitation to participate in this important meeting.

Here are some of the challenges that policy-makers face. At the international level, governments are confronted with the need to reach agreement among over 170 nations on rules that will hold sovereigns accountable for meeting their legally binding greenhouse gas emissions budget limitation and reduction commitments under the Kyoto Protocol on Climate Change,[1] while at the same time affording nations the flexibility to meet those commitments cost-effectively and in a manner

1 The 1997 Kyoto Protocol is a protocol to the 1992 United Nations Framework Convention on Climate Change (UNFCCC). See www.unfccc.de.

that spurs innovation and dissemination of new emissions-reducing technologies and processes and invites broader voluntary participation of other nations.[2]

In the domestic policy arena, the primary challenge that industrialized country governments face is that in economies experiencing robust economic growth, greenhouse gas emissions are rising, and are expected to continue to rise well above the emissions budget levels mandated under the Kyoto Protocol and ancillary agreements.[3] Governments thus face the further challenges of selecting and obtaining political support for policies and measures that can enable governments to comply with their legally binding emissions budgets. The policy instruments considered by governments include taxes and charges; legally binding caps on emissions, with or without emissions trading; voluntary agreements of various types; technology mandates; support for research and development; efficiency standards; specific (per unit output) emissions targets or standards; and various combinations of these and other policies and measures.[4]

Reducing greenhouse gas emissions to meet national targets will present significant challenges for companies and communities as well. They will need to comply with and implement domestic policies and measures to reduce greenhouse gas emissions at the same time that they continue to provide products and services to their customers and constituencies in a cost-effective manner.

But companies and communities face an additional challenge: governmental uncertainty. While the regulatory picture remains uncertain at national and international levels, enterprises and localities need to make near-term decisions about capital stock investments in electricity plants, transportation infrastructure, manufacturing facilities, and the like. Each of these decisions involves consideration of various technology and policy options by entities that may have legal or other obligations to their shareholders and their citizenry to select options that will be cost-effective. For example, failure by a company or community to select a cost-effective option may result in plant closures and job losses.

As local communities and enterprises weigh these decisions in an environment of climate policy uncertainty, they are increasingly interested in a climate policy instrument that can reduce greenhouse gas emissions at the same time that it expands the array of available, cost-effective options: Incentives for Early Action. As described more fully below, Incentives for Early Action programs encourage early emissions reductions in a way that links those reductions to regulatory certainty: The more that a company or community reduces its emissions early, the greater degree of regulatory certainty it will face in the future.

---

2 For a description of outstanding issues in the Kyoto Protocol process, see Goffman et al. (1998).

3 Such ancillary agreements include the burden-sharing arrangements of the European Union, discussed more fully below.

4 For a comparison of various policy instruments at national and international levels, see Petsonk et al. (1998), Wiener (1999) and Petsonk (1999b).

But before explaining what Incentives for Early Action is and how it works, some preliminary words about the Kyoto Protocol and environmental policy alternatives are in order.

The Kyoto Protocol on Climate Change

In 1990, a distinguished international scientific panel, the Intergovernmental Panel on Climate Change (IPCC) issued a report indicating that carbon dioxide, methane, and other greenhouse gases (GHGs), which are being emitted into the atmosphere in ever greater amounts by due to human activities, are residing for decades to centuries in Earth's atmosphere, where they trap heat that would otherwise radiate into space. The result is that our planet is warming more rapidly than at any time since the dawn of modern civilization. Unchecked, anthropogenic emissions of GHGs are expected to contribute to an accelerating warming of the planet, with potentially dangerous interference in the world's climate system.

In 1992, more than 160 governments adopted the United Nations Framework Convention on Climate Change (UNFCCC). This treaty commits its more than 167 Parties to the objective of stabilizing atmospheric concentrations of greenhouse gases (GHGs) at a level that would prevent dangerous anthropogenic interference in the climate system (UNFCCC, Art. 2). The treaty further required industrialized countries to adopt an aim of limiting their GHG emissions to 1990 levels by the year 2000 (UNFCCC, Art. 4.2).

By 1995, it had become apparent that the commitments of the Rio Treaty were not adequate to curb climate change; moreover, many governments were failing to meet a commitment that viewed as voluntary, not mandatory. Accordingly, the first Conference of the Parties (COP) to the Climate Treaty launched a negotiation to adopt a protocol or other legal instrument that would establish legally binding limits on GHG emissions from industrialized nations.

The 1997 Kyoto Protocol establishes those legally binding limits. The Protocol establishes cumulative, five-year, legally-binding caps on the anthropogenic emissions of GHGs by some thirty-nine industrialized nations, with the caps to take effect for the years 2008-2012. The nations and their allowable amounts of emissions are listed in Annex B of the Protocol; hence the appellation, in climate negotiations, "Annex B" nations. The Annex B nations must report their greenhouse gas emissions annually; and must limit their greenhouse gas emissions to the levels established in Annex B (Kyoto Protocol, Art. 3, 5, 7, and Annex B).

The commitments embodied by the Protocol are quite substantial. While previous international legal instruments have embraced legally binding emissions commitments by large groups of nations,[5] this is the first time that a large group of industrialized nations has committed to limit emissions of such a broad range of gases so closely linked with such a broad range of economic activity. Energy, transport, manufacturing, construction, agriculture, forestry – each of these sectors is associated, to varying degrees, with the emission of GHGs.

The Kyoto Protocol allows nations to implement these commitments through an innovative market-based framework. The Protocol allocates to each Annex B nation "parts of assigned amounts" (PAAs) of GHG emissions equal to that nation's allowable GHG emissions under its legally binding cap. It then establishes four flexibility mechanisms through which the Parties may transfer PAAs.

First, Annex B nations may transact PAAs. That is, a nation may transfer PAAs to another nation by subtracting the PAAs from the transferring nation's total assigned amount (sometimes referred to as its 'emissions budget'). In that case, the receiving nation adds the transferred PAAs to its total assigned amount. Emissions trading shall be supplemental to domestic actions for the purpose of meeting emissions limitation commitments (Kyoto Protocol, Art. 3.10, 3.11 and 17).

Second, the Protocol provides that Annex B nations that undertake joint cooperative projects that reduce emissions from the territory of one of the nations may transact parts of assigned amount (PAAs) rendered surplus by those projects. In effect, by agreeing to undertake a joint project under this mechanism, the host government is effecting an allocation of a part of its assigned amount (PAA) to the project as the project's emissions baseline. If the project reduces emissions below the baseline, then the parts of assigned amount (PAAs) that are effectively rendered surplus by the emissions reductions of the project are transferable. The Protocol gives these PAAs a second name, "emissions reduction units" (ERUs). ERUs are identical to PAAs: ERUs transferred in connection with joint projects are, under the Protocol, subtracted from the host nation's emissions budget and added to the receiving (usually, investor) nation's emissions budget. As with emissions trading, the Protocol provides that the acquisition of ERUs shall be supplemental to domestic actions for the purposes of meeting emissions limitation commitments (Kyoto Protocol, Art. 3.10, 3.11 and 6).

Third, the Protocol allows nations that are not members of Annex B to receive certified emissions reduction units (CERs) for projects in their territories that reduce emissions below what would have occurred in the absence of the projects. The Protocol provides that Annex B nations may utilize CERs to meet "part of" Annex B nations' emissions commitments (Kyoto Protocol, Art. 3.12 and 12).

Finally, the Protocol allows Parties to form groups that jointly adopt an emissions limit. Within their group, any two or more Parties may agree to re-allocate parts of assigned amounts amongst themselves. If the group fails to meet its commitment, then each member must meet its commitment under the re-allocation agreement. There is no requirement that group members' intra-group transfers of

5 See, e.g., the Montreal Protocol on the Protection of the Ozone Layer, available at www.unep.org/ozone/

PAAs be supplemental to or part of action to fulfill emissions reduction commitments (Kyoto Protocol, Art. 4). The fifteen member states of the European Union have indicated that they plan to fulfill their Protocol commitments through this reallocation mechanism.

The Kyoto Protocol focuses on the environmental goal of reducing *total*, or *absolute*, GHG emissions. It is environmentally essential that the Kyoto Protocol's framework have this focus, for it is only by reducing total, cumulative GHG emissions that the Protocol's objective – avoiding dangerous climate change – can be achieved.

At the same time, the Protocol invites policy innovation in each of the countries that participate in its framework, because the Protocol does not dictate or prescribe the policy steps that any particular country must adopt. The fact that the Protocol preserves the sovereign right of each participating nation to choose its own implementation policies is critical to the Protocol's acceptance by parliaments and legislatures, and, consequently, to the ultimate environmental success of the Protocol.

Experience indicates that even nations that share common legal frameworks find it extremely difficult to agree on uniform emissions limitation policies. For example, the member states of the European Union have repeatedly tried to reach agreement on a carbon tax, but they have yet to find a tax formula that can be equitably applied across nations with varying economic profiles, at an overall level that will yield the emissions reductions sought. By contrast, within the Kyoto Protocol's legally binding emissions budget framework, nations are free to determine their chosen suite of policies and measures for meeting their emissions targets. It remains to be seen whether this preservation of sovereign prerogatives will be sufficient to achieve entry into force of the Kyoto Protocol.

Under the emissions trading framework established by the Kyoto Protocol, starting in 2008, the Protocol would create a world-wide market for greenhouse gas emissions reductions. In such a market, companies and countries that could make more greenhouse gas reductions than required would be able to earn money by selling them to countries and businesses facing greater difficulty in making their own cuts. Thus, companies will have a positive economic incentive for making extra greenhouse gas emissions reductions.

A similar economic incentive system can be put in place – and put in place quickly – to stimulate businesses to begin making such greenhouse gas reductions prior to 2008. Under this approach, companies that made such reductions would be able to earn greenhouse gas emission reduction allowances that they could save and use for purposes of meeting their mandatory greenhouse gas emissions reduction requirements. They could also sell them to other companies who might need them for the same purpose. In either case, such a program would make greenhouse gas reductions achieved today or any time before 2008 financially valuable to the companies who made such reductions, in just the same way that extra reductions made after 2008 would be valuable in a greenhouse gas emissions trading market after 2008.

Environmental Policy Alternatives: Case Study of the US SO_2 Trading Program

A key reason why the four mechanisms described above were included in the Kyoto Protocol framework is that experience with similar mechanisms at national and international levels indicates that these mechanisms offer the potential of achieving emissions reductions at far lower cost than other mechanisms that might have been adopted. That experience demonstrates that such a framework can tap the creative energies of many different buyers and sellers, encouraging them to engage in an unending search for ways of reducing emissions better, cheaper, faster.

For example, in the United States, a cap-and-trade system for reducing emissions of sulfur dioxide (SO_2), a precursor of acid rain, has produced a striking result: emissions are being reduced faster than the required rate, at costs dramatically lower than those predicted before the start of the program.

Under the program, the US Environmental Protection Agency (EPA) issues allowances to each coal-fired electric utility at the beginning of each year. Utilities must hold allowances equal to their actual emissions. Any utility that emits more than its allowances must purchase allowances from another utility or face steep financial penalties (US$2000 per ton). Any utility that emits less than its allowances may sell remaining allowances to other utilities, or save those remaining allowances for use in a future year, when the emissions cap will be lower.

This "savings" feature has encouraged utilities to compete to reduce emissions ahead of schedule. That is because every "saved" allowance is, in effect, an asset that serves two purposes. First, saved allowances provide each utility with a greater degree of certainty about its ability to manage its regulatory future, since the utility can use the saved allowance to offset future emissions. Second, saved allowances provide a potential future income stream, since the utility can sell the allowances to other utilities that may find it more expensive to reduce emissions in the future.

Figure 1 illustrates the environmental performance of the US sulfur dioxide emissions trading program. Under the program, not only have the utilities reduced their sulfur dioxide emissions to the mandated level, they have reduced these emissions 40% below the mandated level. The program thus demonstrates the power of incentive-based mechanisms to create a context in which competition acts in favor of environmental protection.

The program also channels competitive pressures in favor of reducing the costs of environmental protection. One way of seeing this effect is to look at the price of allowances that are being traded today or saved for future use. Applying conventional discount factors to recent allowance prices while assuming that allowances being banked will be used some time during the decade between 2000 and 2010 suggests compliance costs of about $400 per ton of emissions reduced when the program is fully implemented beginning in year 2000. Consider that even taking account of the cost-savings potential of emissions trading, the US EPA predicted that costs during this same period would be anywhere from $500 to $750 per ton, the result is that the emissions trading market is reducing costs by 20% to 45% from predicted levels. The savings margin is even greater when compared to predicted compliance costs of $1,000 per ton in the absence of emissions trading (Goffman 1998, Ellerman et al. 1997).

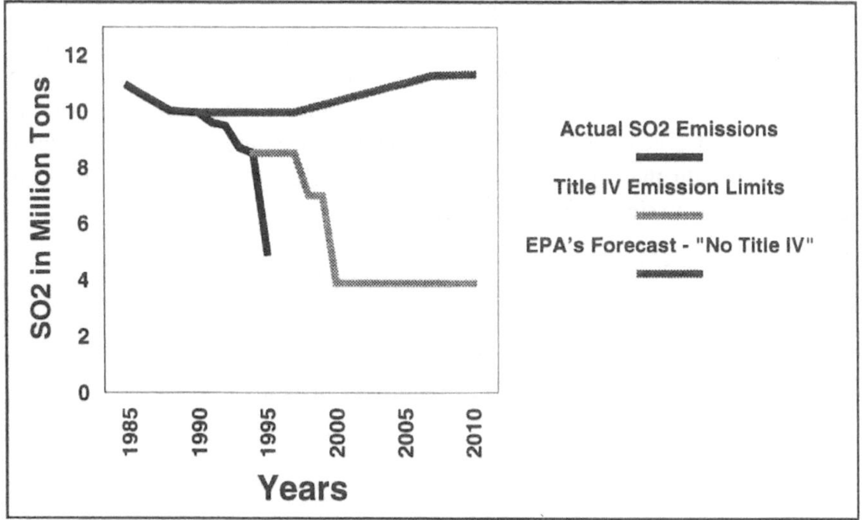

Figure 1: US Sulfur Dioxide Emissions Trading Program Performance

The Climate Policy Challenge: Steep Increases Expected in GHG Emissions

The question logically arises, what lessons can be drawn from the sulfur dioxide emissions trading program described above, and applied to national climate policy?

Figure 2 shows the challenge that many industrialized countries face in meeting the Protocol's greenhouse gas emissions requirements. According to some analyses, US greenhouse gas emissions could rise to more than 30% above 1990 levels if the US continues its robust economic performance on Business-as-Usual (BaU) projections.

Other nations face similar challenges. For example, Sweden's latest set of annual emissions statistics confirm that, if present trends continue, the country is unlikely to meet its Kyoto Protocol target (ENDS Daily, 23 November, 1999).

While Sweden and other member states of the European Union (EU) have each adopted emissions budgets set at 8% below their 1990 levels for the period 2008-2012, the member states have further adopted ancillary agreements to re-allocate these budgets amongst themselves so that some states will need to reduce substantially below the -8% level, while others will have to limit their emissions growth to levels substantially above the -8% level. Under Sweden's re-allocated 'burden-sharing' target, Sweden must limit its emissions to 4% above 1990 levels for the period 2008-2012. So, if all other member states were to meet their burden-sharing targets and Sweden were to follow its current Business-as-Usual (BaU) emissions trajectory, the European Union member states would exceed their joint Kyoto Protocol commitment. If Sweden remains on its current course and follows the upward curve of greenhouse gas emissions shown in Figure 2, then the country will be faced with the prospect of making abrupt changes in order to bring its greenhouse gas emissions down to required levels.

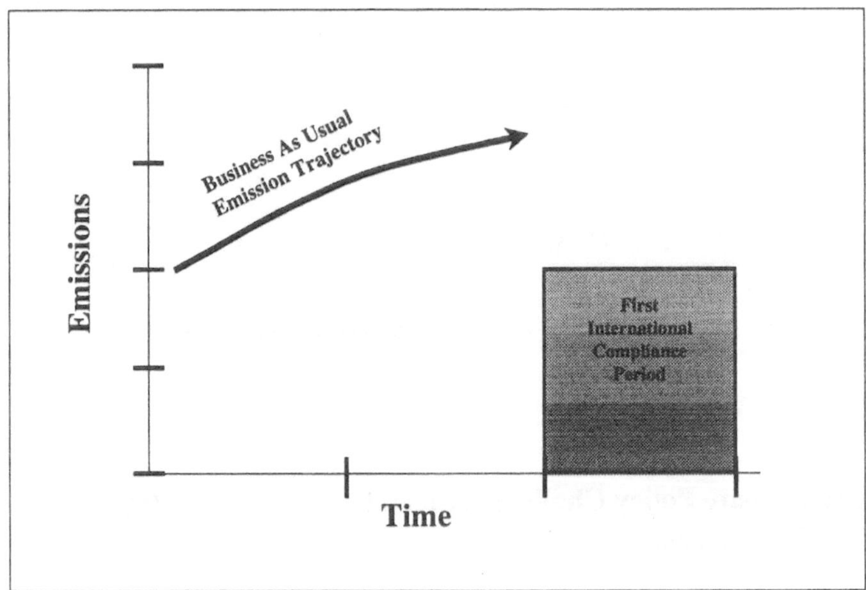

Figure 2: The Climate Policy Challenge

The upward curve poses a serious environmental risk that is likely to be of great concern to environmental policy-makers. Greenhouse gases do their damage by staying in the atmosphere for long periods of time, from decades to centuries. That is why preventing their release in the first place is so important. Because the Protocol's legally binding emissions limits do not begin until 2008, however, the atmosphere faces ten more years of what could be unchecked greenhouse gas emissions increases of the sort shown by the upward curve. Those continued greenhouse gas emissions increases represent that much more warming added to the atmosphere, a warming that could also occur much faster than either the natural world or human societies can tolerate.

It is not only environmental policy-makers, however, that are likely to be concerned about the shape of the curve in Figure 2. Economics policy-makers are also likely to be concerned about the change in emissions trajectory that will be required if nations are to meet their Kyoto targets. It is the abruptness of this change that threatens to inflict the greatest economic pain in the near term, raising political resistance to compliance with the treaty's obligations and limiting the range of choices societies can make in responding to the climate challenge.

Meeting the Climate Policy Challenge: Applying Lessons Learned from the SO_2 Trading Program to Create Incentives for Early Action

The innovative emissions trading approach embraced by the drafters of the Kyoto Protocol offers a way of meeting the climate challenge in the near term. Starting in 2008, the Protocol will create a world-wide market for greenhouse gas emissions reductions. In such a market, countries and companies that reduce greenhouse gas emissions *below required levels* will be able to earn money by selling the surplus emissions reductions they have made to countries and businesses that face greater difficulty in making reductions. This approach creates positive economic incentives for greenhouse gas emissions reductions further than those required.

Using the same basic framework, policy-makers can establish Early Action Incentives programs at national levels and thereby stimulate companies and communities to begin making greenhouse gas reductions *prior to* 2008. Under this framework, companies and local communities that make such early reductions would be able to earn greenhouse gas emission reduction allowances that they could save and use in the future in order to help meet their emissions reduction requirements during the 2008-2012 period. Alternatively, those who earn such allowances could sell them to other companies who might need them for the same purpose. The allowances would be drawn from, or forward-allocated from, the national emissions budget of the country adopting the Incentives for Early Action program.

The signature importance of this type of Incentives for Early Action program is that it would make greenhouse gas reductions achieved today or any time before 2008 financially more valuable to those who make such reductions, in just the same way that reductions below required levels made after 2008 would be valuable in a greenhouse gas emissions trading market after 2008. Thus, the program would have the potential to achieve the same kind of emissions reduction and cost reduction results now being realized in the US acid rain program. The program would achieve this goal by strengthening the economic incentives for companies and local communities to reduce their greenhouse gas emissions *before* legally binding targets take effect.

At the same time, an early reductions incentive program could be entirely voluntary. Paralleling the design of the acid rain program, Figure 3 shows how an Incentives for Early Action program can be designed. Participants who chose to join the program would agree to keep their greenhouse gas emissions at a certain level – somewhere between the levels specified in the Kyoto Protocol and the business-as-usual curve shown in Figure 3.

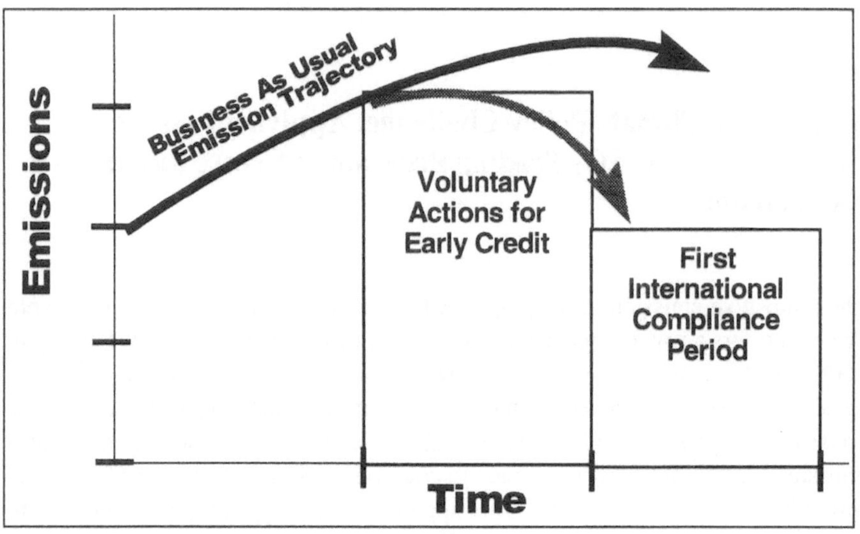

Figure 3: Meeting the Climate Policy Challenge

As shown in the next figure, Figure 4, for any greenhouse gas reductions they made below the specified level, companies and communities that voluntarily choose to participate in the program would receive greenhouse gas emissions reduction allowances, drawn from the total national emissions budget of the nation establishing the policy, for the 2008-2012 period. The recipients could use these allowances to comply with any future, post-2008 obligations. Under such a program, early greenhouse gas reductions – as in the US case of SO_2 reductions between 1995 and 2000 – would have tangible financial value. As a result, companies with opportunities to make greenhouse gas emissions reductions before 2008 would have a compelling financial reason for doing so.

As also shown in Figure 4, an effective Early Action Incentives program would slow, and possibly even reverse, the climb of the upward curve shown in Figure 2. As a result, for nations facing Business as Usual emissions increases above Kyoto Protocol levels, the economic transition to compliance with the Kyoto Protocol's greenhouse gas emissions limits would be smoother and more affordable. Companies that had been able to build up a "bank" of early reduction allowances would have a cost-effective compliance option already on hand when they faced mandatory compliance obligations after 2008. In addition, by giving businesses a direct financial incentive for initiating greenhouse gas investments sooner, an emissions-trading-based early reduction program would ensure that cost-savings innovations were put in place sooner. Consequently, in addition to addressing the short-term economic costs of an abrupt transition to compliance, such a program also would lay the foundation for cost-effective compliance over the long term.

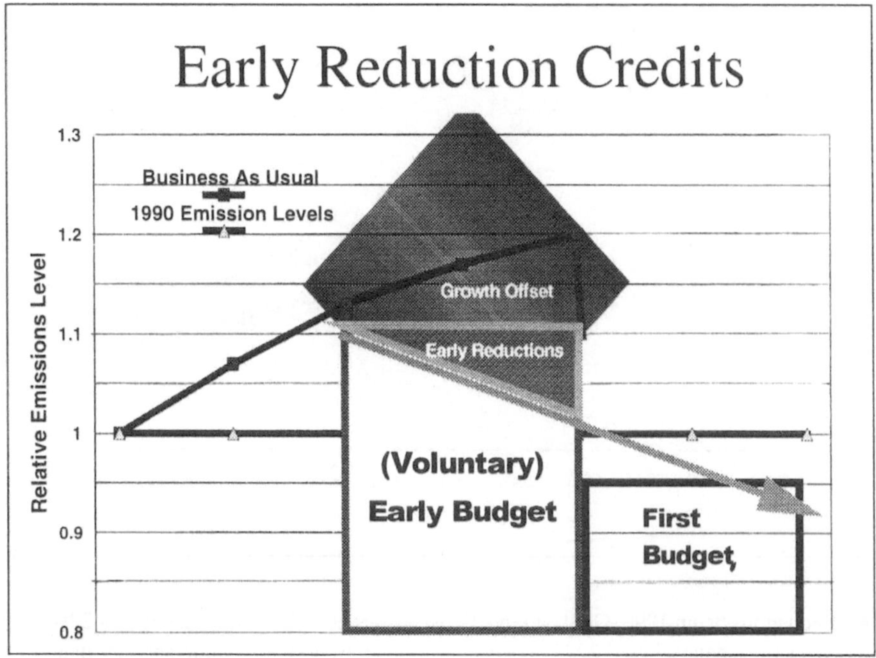

Figure 4: Early Action Incentive Program Policy Framework

At the same time, the environment would benefit directly through the avoidance of additional greenhouse gas emissions prior to 2008, and the discovery and use of environmental innovations could begin that much sooner.

Finally, the climate change issue is likely to increase in its political significance. A well-designed voluntary Early Action Incentives program that provides both economic and environmental benefits has good potential to attract support of both *environmentalists and businesses*, and therefore, of politicians.

Environmentalists are likely to be interested in creating incentives for early action because they recognize the need to begin to bend the Business-as-Usual emissions trajectory of industrialized nations as soon as possible. Many environmentalists seek an emissions limitation and reduction system that, in the early years, will keep open the option to limit global warming to one degree Celsius over the next century, with the goal of preventing dangerous interference in the world's climate system. A concentration of greenhouse gases in the atmosphere of 450 parts per million (ppm) is generally associated with this warming limit. But because greenhouse gases persist in the atmosphere for decades and even centuries, emissions of the gases have to be reduced significantly in order to achieve stable concentrations of the gases.

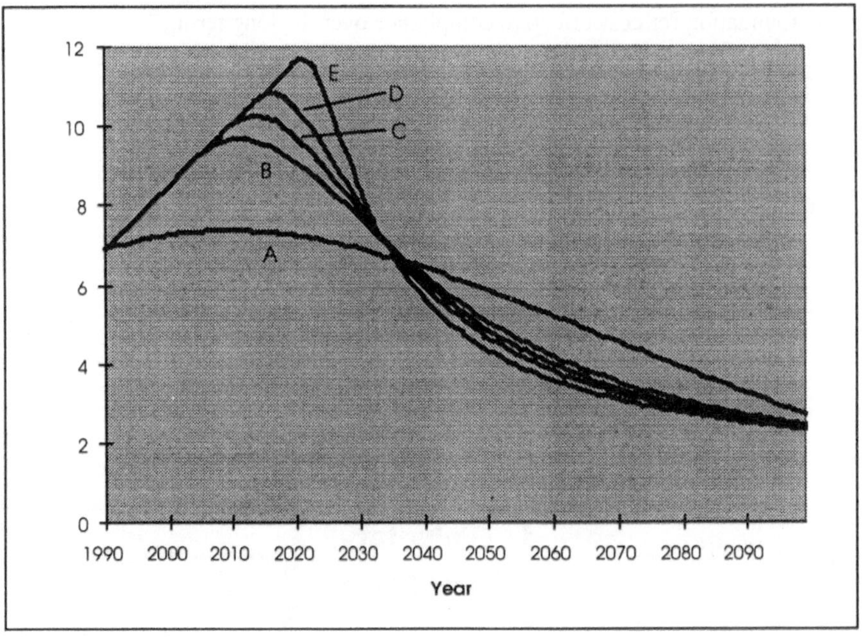

A: Reductions begin 1990. Gradual reductions, at steepest a 2% annual decline by 2080.
B: Reductions delayed until 2005. Steepest decline of 2%/year beginning not later than 2035.
C: Reductions delayed until 2010. Steepest decline of 2.5%/year beginning not later than 2030.
D: Reductions delayed until 2015. Decline of 3.0%/year beginning not later than 2028.
E: Reductions delayed until 2020. Decline of nearly 5%/year beginning not later than 2025.

Figure 5: Emissions Pathways to a Concentration of 450 ppm by 2100

Figure 5 depicts five different emissions paths leading to stabilization at 450 ppm in 2100. The graph shows various emissions paths departing from "Business as Usual" ("BaU"):

- The smoothest ("A") curve is the "S" path identified by the IPCC for reaching 450, and implies reductions below BaU beginning in 1990.
- The next ("B") curve is the curve drawn by Wigley, Richels and Edmonds (WRE), and implies that reductions from BaU are delayed until 2005.

- The "C", "D" and "E" curves show emissions paths that delay reductions below BaU until 2010, 2015, and 2020, respectively.

The curves in Figure 5 demonstrate the consequences of delay by showing the steepness of the reductions required in subsequent years to meet the desired concentration. Steeper, more accelerated reductions are required, and must be undertaken sooner, precisely because initial action has been delayed.

As Figure 5 illustrates, failure to achieve early reductions forecloses the gradual paths, leaving only the paths that require extraordinarily – and perhaps unrealizably – steep reductions in GHG emissions.

Businesses are likely to be interested in the Early Action Incentives framework because the program is voluntary; and it can be established without creating large new bureaucracies. Moreover, such a program would enable those businesses that choose to do so, to manage their emissions and costs in the near term in a way that cushion themselves and their shareholders against future potentially significant economic costs in the implementation of climate policy measures. This is especially true because, from an economic perspective, by creating a more gradual path for emissions reductions over time, early reductions avoid the need for costly, precipitous changes in energy and infrastructure.

A Note on Program Design

As policy-makers move forward to consider Early Action Incentive Programs, each nation will need to develop a program within the basic framework that addresses its specific national circumstances. While the framework provides the context, the particular ways in which national programs fit into that context will depend on a range of choices open to policy-makers. So, for example, while policy-makers will need to take care that any allowances that are allocated to those who undertake early action are drawn from the national emissions budget, the extent of early reductions needed in order to earn the emissions allowance allocation is a matter for national decision-making. Similarly, each nation may wish to develop its own approaches to addressing early allocations to industries that are growing; to early reducers who enhance carbon sequestration in forests and agriculture; and to answering a number of other program design questions. As long as the basic framework described above is adhered to, there is considerable flexibility available to countries to experiment with and learn from particular program design.[6]

---

6 By and large, countries may experiment with these different program designs without fear that their programs will raise trade issues. See generally Petsonk (1999b) and sources cited therein for both this and contrasting views.

Developing Countries and Incentives for Early Action

One of the benefits of a well-organized conference like this one is the opportunity for cross-pollination of policy ideas. At this conference, one of the panelists, Dr. Ewers, President of the Free University of Berlin, put forward an innovative policy proposal for recognizing and rewarding the participation of developing countries in climate protection efforts. He suggested that those developing countries that move early to adopt national emissions budgets might be awarded an incentive – a first mover advantage, in the form of an increment of emissions allowances greater than that which they might otherwise receive.

This interesting idea merits serious attention by climate policy-makers. While it could be argued that such an approach would simply award so-called 'hot air' to developing nations, that problem can be avoided by ensuring that the total allocation such early movers receive does not push the world off the emissions pathways depicted in Figure 5. If this constraint is met, then an emissions 'premium' for first movers in the developing world could, if paired with the Early Action Incentive program framework described in this paper, provide a very important mechanism for encouraging early investment in building climate-sustainable infrastructure in developing nations at precisely the time that the Business-as-Usual emissions path of those nations is expected to grow steeply.

Conclusion

The Incentives for Early Action framework described above is intended to serve two purposes. First, it offers an illustration of how important the emissions trading policy tool is in ensuring that compliance with the Kyoto Protocol is economically affordable. Second, the Incentives for Early Action framework strategy provides a valuable means of achieving critical early emissions reductions while smoothing the economic transition that the climate challenge presents.

References

Ellerman, D.A., Schmalensee, R., Joskow, P., Montero, J.P. & Bailey, E. (1997), *Sulfur Dioxide Emissions Trading Under Title IV of the 1990 Clean Air Act Amendments: Evaluation of Compliance Costs and Allowance Market Performance.* Center for Energy and Environmental Policy Research, Massachusetts Institute of Technology, Cambridge, MA.

Goffman, J. (1998), *Testimony before the Committee on Science of the U.S. House of Representatives*, February 4, 1998.

Goffman, J., Dudek, D., Oppenheimer, M., Petsonk, A. & Wade, S. (1998), *Cooperative Mechanisms Under the Kyoto Protocol: The Path Forward.* EDF, June 1998, available at www.edf.org.

Petsonk, A. (1999a), *Taxes and Trading: The Context for Climate Policy Instruments.* Heinrich Böll Foundation.

Petsonk, A. (1999b), *The Kyoto-Protocol and the WTO: Integrating Greenhouse Gas Emissions Allowance Trading Into the Global Marketplace.* Duke Environmental Law and Policy Forum (forthcoming).

Petsonk, A., Dudek, D. & Goffman, J. (1998), *Market Mechanisms & Global Climate Change: An Analysis of Policy Instruments.* Report prepared for the Trans-Atlantic Dialogues on Market Mechanisms, The Pew Center on Global Climate Change, October 1998, available at www.pewclimate.org .

Wiener, J. (1999), Global Environmental Regulation: Instrument Choice in Legal Context, 108 *Yale Law Journal* **677**.

A Proposal for Credible Early Action in US Climate Policy

Raymond Kopp, Richard Morgenstern, William Pizer and Michael Toman [1]

Resources for the Future, 1616 P Street NW, Washington, DC 20036, U.S.A., e-mail: kopp@rff.org.

As international negotiations on climate change continue, momentum is building for domestic "early action" to begin reducing US emissions of greenhouse gases in the nearer term. With unopposed Senate ratification of the Framework Convention on Climate Change (FCCC) in 1992, the US and other Parties are already committed to adopt policies and measures to limit these emissions.

While the mostly voluntary measures undertaken so far have had some positive effects, the US and virtually all industrialized nations remain on a trajectory of significant emission growth. In particular, the limited success of these measures suggests that businesses and individuals are abstaining from significant new investments in emission reducing capital and research. This pattern will undoubtedly continue until there is greater clarity about both domestic and international policies.

Current proposals for early action continue to emphasize voluntary measures that do not provide clear and substantive incentives for credible reductions. Senate Bill S.547,[2] for example, would credit early, voluntary emission reductions with an equivalent volume of emission rights under an unspecified future permitting scheme. In theory, a business that plans to undertake reduction activities in the future, once a permit program is in place, would have a similar incentive to undertake such activities much sooner thanks to early crediting.

---

1 Reprinted from Weathervane, August 1999, http://www.weathervane.rff.org/features/feature060.html.

2 http://www.weathervane.rff.org/refdocs/s547is.txt.pdf.

From the nation's perspective, however, this proposal risks distributing too many credits for questionable early reductions. The only way to reduce this risk is to thoroughly examine each project and evaluate the true reductions incurred – a cumbersome and potentially expensive administrative process. Further, the proposal still connects the incentive for early action to the future worth of emission rights under an uncertain permit scheme, a speculative value that may fail to motivate sufficient activity.

In lieu of new voluntary measures we propose that the US initiate a domestic tradable permit program with three key features: broad coverage, a modest target, and equitable burden-sharing.
We believe a well-designed domestic trading program for carbon dioxide could be established by 2002. Such a policy would create genuine incentives to look for emission reductions now and to develop new low-emission technologies for the future. These key features guarantee that the reductions are undertaken in the most efficient possible manner, that the cost of the reductions remain economically acceptable, and that the burden to individuals is both limited and equitably shared.

We propose that the program be administered "upstream" to obtain the broadest possible coverage.
Broad coverage guarantees that all sources of carbon dioxide emissions face the same incentive to cut back and therefore aggregate reductions are obtained at the lowest possible cost. This should be true regardless of whether those reductions occur among electric utilities, in the transportation sector, or elsewhere. In an upstream program, we focus on domestic energy producers (and importers) in order to obtain this broad coverage at the lowest possible administration and monitoring cost.

In particular, we would require energy producers to obtain permits equivalent to the volume of carbon dioxide eventually released by the fuels they sell.
By collecting permits at the mine mouth for coal, the refinery gate for crude oil, and at the initial point of distribution for natural gas, virtually all domestic emissions are covered by roughly two thousand collection points. This is then augmented by a permit requirement on imported fuels along with exemptions for non-combustion use or export. The key point is that this approach provides the same incentives as a more complex, more expensive, and less comprehensive downstream program focused on end-users

The program should be broadened as quickly as possible to include other greenhouse gases, sinks, and international joint implementation projects.
These sources represent additional opportunities for equally effective and potentially low-cost reductions in net greenhouse gas emissions. Yet, they also pose greater challenges to monitoring and enforcement. We believe that rules and regulations for crediting carbon sinks and joint implementation emission reduction projects as well as controlling other greenhouse gases – Methane (CH_4), Nitrous oxide (N_2O), Hydrofluorocarbons (HFCs), Perfluorocarbons (PFCs), and Sulfur

hexaflouride (SF_6) – can be developed by 2004 or sooner. To the extent practicable, these regulations should permit trading among all gases, sinks and joint implementation projects, and should be consistent with internationally accepted definitions.

An essential element of this proposal is that it remains modest.
This modesty reflects the desire to proceed gradually while we undertake further research into climate change consequences, while the capital stock (both human and physical) adjusts to new incentives, and while other countries remain undecided about their own courses of action. Yet, one of the potential problems associated with any permit trading scheme is the large uncertainty concerning future permit prices, and by extension the compliance cost of the program.

Therefore, we propose capping the price of permits in order to prevent the program from becoming too expensive.
With a price ceiling – or safety valve – the government provides unlimited additional, above-target permits at a specified price. As long as the price remains below the ceiling, emissions are strictly limited to the number of permits in the initial distribution – the target. If the permit price reaches the price ceiling, extra permits are offered for sale and emissions are allowed to rise in order to contain compliance costs.

This ceiling price should therefore reflect the maximal burden that society is willing to bear in order to reduce a ton of carbon dioxide.
We propose an initial ceiling price of $25/ton carbon in 2002 which rises by 7 percent each year (above inflation) through 2007. In 2002, this ceiling price is equivalent to a six cent rise in gasoline prices – less than one-third the price decline experienced in the past year. The equivalent rise in electricity prices is one-half cent per kilowatt hour. By beginning with a low ceiling price that rises over time, we provide a softer transition to these new incentives and send a signal that energy prices are likely to rise, not fall, in the future. This signal strengthens the incentive for technological innovation that is essential for future emission reductions.

Based on the ceiling price we can then determine an appropriate target for annual emissions.
Current analyses suggest that these prices would yield annual emissions of roughly 1,460 million tons of carbon (MtC) through 2007. We propose an initial distribution of permits equal to this estimated volume. This was the actual emission level in 1996 and represents both a ten percent reduction from the forecast emission level in 2002 (1,621 MtC) and a ten percent increase over the 1990 emission level (1,346 MtC). With business-as-usual (BAU) forecasts indicating 1.3 percent annual growth in emissions, 1,460 MtC represents a 20 percent reduction from BAU by 2008.

Although permits will be freely tradable, we propose that ordinary permits expire after two years and permits sold at the ceiling price expire after one year.
Since future climate change goals are uncertain, we need to preserve the option of lowering emission targets in the future if it becomes warranted. This option could be thwarted if there is an excessive accumulation of valid but unused permits in the system. To avoid this risk we place limits on permit banking. With large, sophisticated markets for carbon, firms will be able to efficiently sell excess permits and purchase options on future permits, thereby eliminating the need for long-term of banking.

The final cornerstone of our proposal is that the program be equitable: therefore permits must be auctioned and the revenues generated through permit sales must be returned to households.
Since any carbon reduction program will raise the cost of carbon-based fossil fuels, households will bear a large direct burden in the form of higher household energy prices as well as an indirect burden in the form of higher prices for those goods and services with a high energy content. If the program is going to equitable – and indeed viable – it must compensate households for these price increases. Thus, we propose quarterly permit auctions with 75 percent of all proceeds in the first year funding a direct payment to all US households based on legal residency. In order to address special hardships, the remaining 25 percent would be given to states based on energy use by low income households and the vulnerability of industry (both owners and employees) to increased energy costs. In subsequent years, we propose reducing the share that accrues to states by 2.5 percent annually as these special hardships are successfully addressed. Revenue generated by extra permit sales at the specified ceiling price would follow the same allocation as auction revenues.

By establishing a credible early action program, the United States will learn a great deal about the opportunities and costs associated with greenhouse gas reductions – information which can enhance our ability to make prudent decisions. We also expect that similar motivations will lead other nations to adopt polices and measures designed to reduce emissions. Accordingly, it is critical that a major evaluation of the US and other nations' programs be conducted during the first five years of implementation. The evaluation should consider the environmental, economic, and social performance of the program. On the basis of that evaluation, the actions of other countries, and the status of the Kyoto negotiations, the United States can then decide how best to proceed in the future.